The Invisible Language of Nature

Ivan G. Ivanov

The Invisible Language of Nature

How Plants and Animals Use Chemical Signals

Ivan G. Ivanov
Institute of Molecular Biology "Rumen Tsanev"
Bulgarian Academy of Sciences
Sofia, Bulgaria

ISBN 978-3-662-73301-1 ISBN 978-3-662-73302-8 (eBook)
https://doi.org/10.1007/978-3-662-73302-8

This book is a translation of the original German edition "Die unsichtbare Sprache der Natur" by Ivan G. Ivanov, published by Springer-Verlag GmbH, DE in 2025. The translation was done with the help of an artificial intelligence machine translation tool. A subsequent human revision was done primarily in terms of content, so that the book will read stylistically differently from a conventional translation. Springer Nature works continuously to further the development of tools for the production of books and on the related technologies to support the authors.

This Springer imprint is published by the registered company Springer-Verlag GmbH, DE, part of Springer Nature.
The registered company address is: Heidelberger Platz 3, 14197 Berlin, Germany

Contents

Introduction

Every living cell, every living organism is a complex chemical laboratory, whose perfection has already earned the reverence of more than one chemist. Thousands of chemical reactions take place in the cell every minute, in which complex syntheses of important substances occur at the expense of the breakdown of other, simpler compounds. A living organism is inseparably connected with its environment. It constantly exchanges chemical substances with it. Mineral salts and nutrients are absorbed, and the products of its life processes are excreted. Since our environment consists of both inanimate matter and living organisms, every inhabitant of our planet is directly or indirectly connected with both the inanimate and the living world. These complex relationships were first recognized by the great French chemist Antoine Lavoisier (1743–1794; discoverer of oxygen and silicon). In his historic work "The Circulation of Elements on the Earth's Surface," he writes: "Plants obtain the substances necessary for their life from the surrounding air, water, and the entire inanimate nature. Animals feed either on plants or on animals that feed on plants, so that the substances of which their organisms are composed ultimately come from the air and the mineral kingdom. Finally, through fermentation, decay, and combustion, all these substances taken from plants and animals are constantly returned to the air and the mineral kingdom."

Living organisms are interconnected in complex food chains, and the disruption of a single link leads to an imbalance in the entire chain. But the connection between them is by no means limited to nutrition. They depend on each other in a much broader sense. Microorganisms, for example, are

I. G. Ivanov, *The Invisible Language of Nature*,
https://doi.org/10.1007/978-3-662-73302-8_1

needed as nature's sanitarians in the decomposition of the carcasses of dead animals and plants. This frees up habitat for new generations. In addition, many microorganisms live in symbiosis with animals and plants, supplying them with important biologically active substances that they themselves cannot produce. Some lower organisms such as fungi, lichens, mosses, etc., are soil formers. Their appearance usually precedes that of higher plants. Insects are needed by plants for pollination and by these for nourishment. The unisexual individuals are mutually indispensable for the survival of the species, and so on. For every human, the other inhabitants of the earth can be roughly divided into useful and harmful. Accordingly, their attitude toward them also differs.

In the course of evolution, living beings have developed and established a multitude of means and ways to communicate with their own kind as well as to defend themselves against enemies. One of the most widespread and common means of communication among all living beings is chemistry. Even the oldest inhabitants of the earth produced substances that were not directly necessary for the functioning of their own organism, but for their interaction with other organisms in the same habitat. For some, they are attractants (with an attractive effect), for others, repellents (with a repellent effect), and for still others, poisons. With the emergence of higher organisms, especially animals, specialized organs for the production and reception of such substances developed.

On the basis of attractants and repellents, a highly sensitive and efficient form of communication has developed, through which individuals of the same species can exchange vital information. The released chemical substances are beneficial to both the producing and the receiving organism. But there are also substances that can severely disrupt or completely inhibit the life processes of the receiving organisms, leading to their demise. Such substances benefit only the producer and harm the affected individual. Thus arose the chemical weapon of defense and attack, which in some species has reached an incredible level of perfection.

The fact that chemical communication can be observed in both lower and higher organisms shows that the chemical mode of communication has proven itself over time and has contributed to the preservation of many biological species. Herodotus wrote: “In the long run only the most probable occurs,” and the chemical ecologist Michel Barbie says: “In the long run, only the most stable and necessary remains.” In the sense of the latter, we can say that the chemical means of communication in living nature, including means of defense and attack, are vital for the organisms that produce them. For most of them, they are not a luxury, but a necessity for survival.

For all inhabitants of the earth, except humans, the question is not how they can live better, but how they can live at all. Thousands of plant and animal species owe their existence to the perfection of their chemical laboratories.

While poisonous plants and animals have been known to humans since antiquity, the possibility of exchanging information via chemical compounds was only discovered in the last century. At that time, pheromones were discovered—an extremely important class of biologically active compounds. With them, individuals of the same species share important information about themselves, food sources, impending dangers, the maintenance of order and harmony in their communities, the control of population density, and so on. Due to their low concentration in the living environment, studying their chemical nature and the mechanisms of their action was no easy task. Only with the advent of highly sensitive instrumental methods for structural chemical and physiological investigations could successes in this area be achieved.

The aim of this book is to introduce the reader to the fundamentals of chemical communication between living organisms, plants, and animals. The significance of chemical signals for the life of lower organisms is treated only cursorily, mainly to emphasize the universal character of the chemical language in living nature.

The greater part of the book is devoted to the chemical substances that serve the exchange of information between organisms. Less space is given to chemical means of defense and attack, i.e., chemical weapons of biogenic origin. In doing so, we have been guided by the fact that much has already been written about biological poisons, while information about pheromones and musk is rather sparse. Of the poisonous animals, we have selected those whose chemical weapons are relatively well researched and of interest to humans.

Chemical Communication in Nature

> Chemical communication is the oldest form of interaction between organisms. It arose at the dawn of life on Earth and was perfected by natural selection.
>
> Prof. Y. D. Kirshenblat

Why is chemical communication between organisms so ancient? To answer this question, one needs a bit of imagination and a wealth of scientific facts. Let us imagine our planet 3–4 billion years ago. At that time, it was shaken by powerful volcanoes, meteorite impacts, and intense tectonic activity. Gases escaped from huge fissures and craters on its surface and became part of its atmosphere. Since there was no ozone layer, the barren rocks were exposed to much more intense solar radiation than today. Any living being that happened to arrive on Earth back then would have been instantly killed by the intense radiation and the toxic primordial atmosphere. Radical changes had to occur on Earth before life could arise. The presence of large bodies of water was also necessary. Water collected in low-lying areas, forming lakes, which in turn merged into seas and oceans. How many years it took for the world's oceans to fill is unknown. However, it is likely that the water level 2–3 billion years ago was roughly as high as it is today. With the formation of large bodies of water, the water cycle on Earth began. Water evaporated, saturated the atmosphere with water vapor, and returned to the oceans after condensation. The presence of water vapor in the atmosphere is an important factor for the emergence of life, as it, together with the ozone of the ozone layer, absorbs deadly ultraviolet rays.

I. G. Ivanov, *The Invisible Language of Nature*,
https://doi.org/10.1007/978-3-662-73302-8_2

The Earth's first atmosphere contained carbon dioxide, water vapor, nitrogen, ammonia, and other gases. Their interactions also gave rise to the first simple organic compounds. Rain carried them into the ocean, where they were transformed into more complex molecules—prototypes of the carbohydrates, proteins, and nucleic acids we know today. Over millions of years, their concentration increased, leading to the formation of the "primordial soup"—the environment in which life is thought to have originated. Exactly how this happened is not entirely clear and will probably never be fully resolved. According to theories of spontaneous generation, the first living organisms were the product of an increasingly complex organization of the planet's organic matter. The panspermia hypothesis, on the other hand, suggests that simple life forms traveled vast distances through the universe and thus brought the beginnings of life to Earth. In this case, the primordial soup was simply a lavish meal for the visitors to planet Earth (Fig. 1).

Fig. 1 Planet Earth as seen by ancient geographers and philosophers. (© katatonia82/ Getty Images/iStock)

We also do not know exactly what the first inhabitants of Earth looked like. They may not have resembled the living beings we know today. Most likely, they were similar to bacteria. Regardless of their biological characteristics, we can say with great certainty that they were unicellular heterotrophic organisms, i.e., they lived off ready-made, abiotically synthesized organic substances. The latter were broken down into simpler compounds, and the energy released from them was used to meet the needs of life. However, food was unevenly distributed across the Earth's surface, and the first inhabitants, primitive as they were, had to search for and find sufficient food sources and distinguish edible from toxic chemical compounds.

Therefore, from the very beginning of life, the first cells must have "understood" the significance of chemical messages from their environment. Their perception is nothing other than chemoreception, and their response—chemotaxis. Depending on whether the chemical source attracts or repels the organism, chemotaxis can be positive or negative. Only cells capable of correctly interpreting chemical signals could survive and evolve.

As the number of living organisms increased, the primordial soup was gradually depleted. To avoid dying, the inhabitants of Earth faced the dilemma of either having to absorb carbon dioxide and nitrogen from the environment or feeding on their competitors and conspecifics. Unsurprisingly, both approaches had their "followers." The first method led to autotrophic photosynthetic organisms such as cyanobacteria and green plants, the second to heterotrophic organisms. The latter probably invented the first chemical weapons, the antibiotics (Greek ἀντί- anti- "against" and βίος- bios "life"), with which they destroyed their competitors. This, in turn, prompted the peaceful autotrophic organisms to "seek" a means of protection against their enemies. To gain more power, they too turned to chemistry and began to produce phytoncides. Later-emerging animals also benefited from the services of chemistry. As their populations grew, they developed sophisticated chemical defenses, which in some cases were used not only for protection but also for attack.

With the appearance of the first unicellular organisms, the complex process of speciation began. The emergence of different biological species led to the need to recognize individuals of the same species. Today we know that the periodic renewal of genetic material through so-called conjugation is necessary for stable species maintenance in bacteria (Fig. 2). This occurs through the contact of two cells of the same species, resulting in the formation of an intercellular bridge. Through this, genetic material is transferred from one cell to the other. This process is similar to the sexual process in higher organisms.

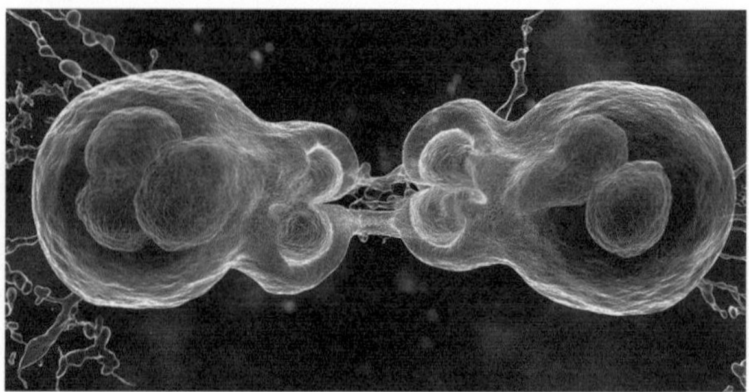

Fig. 2 Conjugation in bacteria. (© Andres Mejia/Generated with AI/Stock.adobe.com)

For conjugation to occur, the two cells must find and recognize each other in the vast space available to them. Since they have no other senses, the only way to recognize each other is through the exchange of chemical signals. To this end, bacteria and later higher organisms began to secrete species-specific substances that are recognized only by individuals of the same species. By moving toward the increasing concentration (gradient) of the secreted substance, they could unmistakably identify the source of the chemical signal. This phenomenon is known as positive chemotaxis.

Over the course of evolution, multicellular organisms gradually emerged. Their precursors were colonies of unicellular organisms, which still exist today. To maintain a colony of conspecifics, two types of messenger substances began to be produced: one for communication between cells within the colony, the other for communication between individual colonies. The former can be regarded as prototypes of hormones, the latter as prototypes of pheromones.

Although highly developed sensory communication systems such as visual and auditory systems ("eyes" and "ears") later evolved in higher organisms, chemical communication not only did not lose its importance but in many cases was further refined. Over the course of evolution, specialized secretory organs developed—so-called endocrine glands (secretion into the organism), as well as those with secretion into the environment (exocrine glands—Fig. 3). In addition, organs for the perception of chemical signals—olfactory organs—emerged. The information content of chemical signals became so sophisticated that animals living today can recognize not only their conspecifics but also their enemies by scent.

Chemical signaling is the basis for organization and maintenance of order in some of the most structured communities on our planet—the colonies of social insects (bees, ants, termites, etc.).

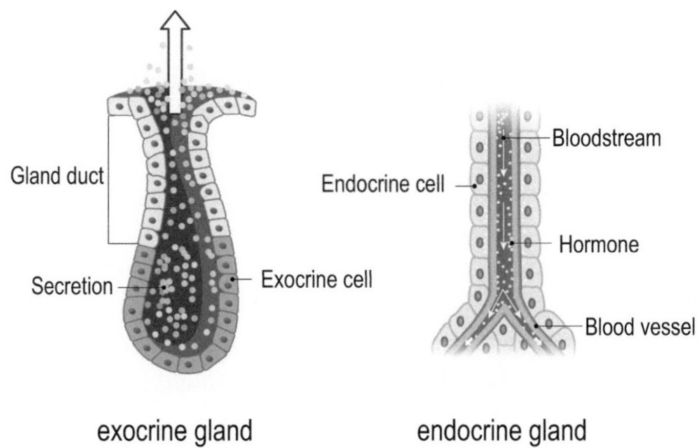

Fig. 3 Diagram of an exocrine and endocrine gland. (© ttsz/Getty Images/iStock)

Evolution has brought some animals into close dependence on plants and vice versa. From the plant's perspective, animals that use them as food are either harmful (if they destroy them) or beneficial (if they help them change/reproduce). This results in their different attitudes toward the two groups of animals. They repel the harmful and attract the beneficial. Chemical signals also underlie the complex relationships between plants and animals. The plant is able to produce substances that are attractive to some, repellent or toxic to others. In other words, the universal language of chemistry is understandable even to representatives of different kingdoms of living nature.

Living cells thus secrete chemicals for different purposes. In general, we can say that endocrine glands produce signaling substances (hormones) intended for communication with the cells of an organism, while exocrine glands are responsible for communication with cells of another organism. The latter make an important contribution to the maintenance of the population and the species. From a scientific perspective, the two types of glands have not been studied with the same care and depth. While much is known about endocrine glands and their products, and libraries are filled with original and review literature, our knowledge of exocrine glands and their products is much more modest.

The substances produced by exocrine glands are called semiochemicals, a term derived from the Greek word *σημεῖον* (semeion), meaning signal. In other words, semiochemicals are signaling substances that serve the exchange of vital information between organisms. Depending on whether they originate from the same or a different species, semiochemicals are intraspecific or interspecific. Pheromones belong to the former, allomones, kairomones, and synomones to the latter. The difference between allomones and kairomones is that the former benefit the producing organism, while the latter benefit the receiving organism. Allomones include various substances for defense against enemies, antibiotics, poisons, antidotes, baits, etc., while kairomones include substances that attract organisms to food sources, danger signals, growth factors, or stimulants, etc. Synomones, in turn, benefit both the producer and the receiver. These include substances secreted by certain orchid species that mimic the female sex pheromones of specific flying insects and stimulate their males to visit and pollinate the producing plants.

Representatives of intraspecific semiochemicals are pheromones. The term was first introduced by P. Carlsson and M. Lüscher in 1959 and is derived from the Greek words φερο (fero, to carry) and ορμόνη (hormone). Depending on their purpose, pheromones are classified as sex, social, alarm, and trail pheromones. M. Florkin was one of the founders of biosemiotics (life as a biological process of signs and communication) and included all substances that fulfill one of the aforementioned functions. Kirshenblat called all biologically active substances released by living organisms into the external environment telergones. He proposes calling intraspecific substances homotelergones and those with interspecific effects heterotelergones.

Given the comprehensive and complex nature of chemical interactions in nature, the question naturally arises as to which science should study chemical signaling in the biosphere. The investigation of the material nature of chemical signals, i.e., the molecular structure of the secreted biologically active substances, is of course the domain of chemistry, and in particular natural products chemistry. Their biosynthesis is studied by biochemistry in the field of secondary metabolism. The mechanisms of action of semiochemicals are in turn studied by teams of biochemists, biophysicists, physiologists, etc. Since their effects are associated with specific behavioral responses, the study of their influence on animal behavior falls within the scope of ethology (behavioral research). And because the effects of signaling molecules extend beyond individual species, their impact on animal communities is studied by ecology.

In recent years, the view has gained ground that the complex study of substances produced by living organisms and released into the external environment should be the subject of its own scientific discipline, a branch of ecology. Years ago, this science was called "chemical ecology." This imprecise name was proposed on the grounds that the chemical substances in question are released into the external environment and ecology is the science that studies the relationships between organism and environment. Although the term chemical ecology is still in use, it is not the most appropriate. By analogy, the branch of physics dealing with biogenic sound and light signals would have to be called "physical ecology," which is only partially correct. Perhaps the name will emerge once the subject of the new scientific field has become established. The French scientist Michel Barbier believes that the term "biocoenotic ecology" is more appropriate, as it encompasses both intraspecific and interspecific relationships between organisms.

Chemical ecology hopes to uncover many previously unknown cause-and-effect relationships in the living world as well as between living and non-living nature. Deciphering the meaning of chemical signals will enable humans to intervene more effectively and in a more environmentally friendly way in the life of ecosystems, using much subtler and more eco-friendly means. Unlike the methods used so far, these will not have collateral destructive effects.

The Chemical Messages of Insects: The Molecules of Love

Countless short stories and novels are dedicated to love. The feelings of love move the hand of the artist who creates the image of the beloved woman, the poet who immortalizes his muse, and the composer who pours his emotions into music. Everyone finds a suitable way to express their feelings according to their talent. But love is not only an attribute of *Homo sapiens,* of humans. It is also inherent in representatives of the animal kingdom, even though they cannot write, speak, or compose. Much has been written about their courtship rituals. The curious reader can find almost everything of interest in the extensive popular science literature. In the following, we will focus on a special and lesser-known form of love communication—courtship through chemical signals.

Sentimental souls may ask upon reading these lines: What does chemistry have to do with love? They should not judge too hastily. Let us, for a moment, put ourselves in the position of the billions of inhabitants of our planet who, unlike us, do not possess reason and perceive the world in a completely different way. We should not forget that our perception of reality is more or less subjective. For example, while nature is colorful for a sighted person with normal vision, it is black and white for the color-blind; for the hearing, the world is loud, but for the deaf, it is silent. Our perception of the world is based on information that reaches our brain through the five basic senses—sight, hearing, smell, taste, and touch. For humans, these senses are not equally important and are in a balanced relationship with each other. About 80% of external information is received through sight and hearing,

I. G. Ivanov, *The Invisible Language of Nature*,
https://doi.org/10.1007/978-3-662-73302-8_3

i.e., through the eyes and ears. In animals, however, the various senses have different significance for information intake depending on their way of life. For example, a dog perceives 80% of the world around it through its sense of smell. G. Dekar, author of works on animal sensory systems, writes: "If a dog were to write a book about the sense organs, the longest chapter would undoubtedly be devoted to the sense of smell." The sense of smell is no less, and even more, important for other animals such as insects, fish, rabbits, etc. In other words: while for us and the primates the world consists of 80% sound and light, for many living beings it consists of 80% chemical signals. And then, logically, the question arises: Why shouldn't the courtship dance of animals with developed vision or the love song of animals with developed hearing be replaced by chemical love messages in animals with a highly developed sense of smell?

The first evidence for the existence of love molecules can be found in the work of the famous French entomologist Jean-Henri Fabre (Fig. 1). In 1904, he placed a freshly emerged female moth of the species *Saturnia pyri* (giant

Fig. 1 Jean Fabre (1823–1915). (© opale.photo/Darchivio/picture alliance)

Fig. 2 *Saturnia pyri*. (© Alexey Protasov/Getty Images/iStock)

peacock moth, Fig. 2) in a box and covered it with a gauze blanket. That very evening, the scientist observed that the young female was surrounded by about 40 suitors of the same species. This experiment was repeated on the following days with identical results. The males found the female even when odorants such as naphthalene, essential oils, etc., were sprayed around her. The female's attraction was so great that the suitors entered through the professor's office windows or down the chimney if the windows were closed.

Even then, Fabre assumed that the female moth emitted chemical signals that the males perceived through their sense of smell. Supporting his hypothesis was the fact that the empty box continued to attract males even after the female was removed. The attraction was greater the rougher and more porous the surface of the container. Cardboard, clay, and sand had the greatest effect, while marble and metal surfaces were less effective.

Similar experiments were conducted with the silkworm *Bombyx mori* (Fig. 3b) by Adolf Butenandt (Fig. 3a), a German chemist and Nobel laureate in Chemistry in 1939. He confined female silkworms in specially prepared cages and allowed marked males to fly to them. There were recorded cases in which males found females from a distance of 11 km.

Butenandt attempted to isolate the love scent of *Bombyx mori* by extracting the abdomens of 7000 moths with benzene. In this way, he obtained 1.5 g of a waxy substance that exerted a strong attraction on the males. Filter paper soaked with this extract excited the moths so much that they attempted to copulate with the paper itself. Since the benzene extract contained many other ballast substances, its quantity was insufficient to isolate an active ingredient in pure form. Due to the small amount of the substance, the scientist was only able to demonstrate that it was chemically a type of alcohol.

Butenandt's experiments lasted 20 years. Each time, he increased the amount of extracted material and improved the purity of the active ingredient. In the meantime, analytical methods were also improved. Spectral

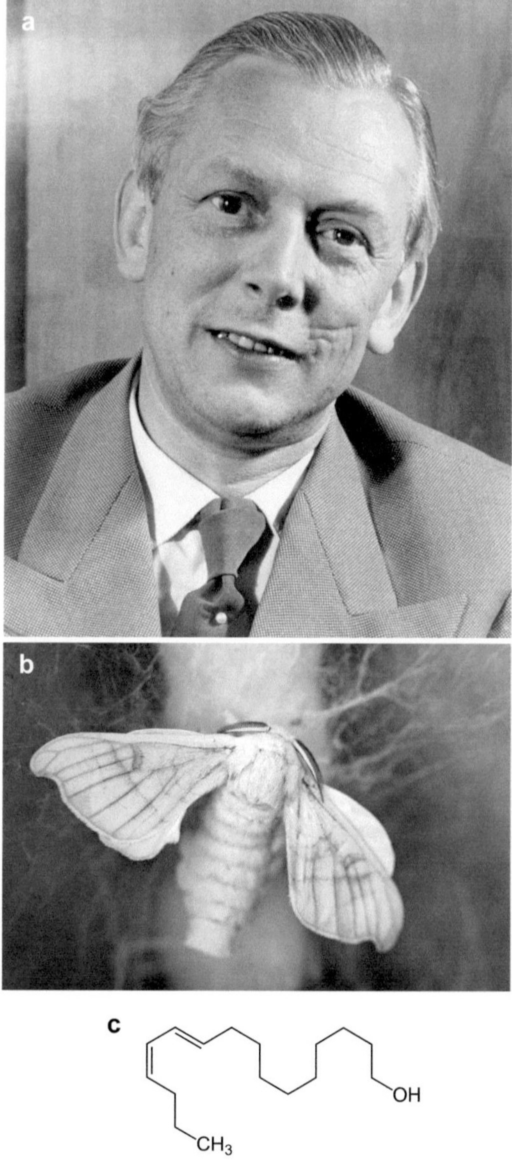

Fig. 3 **a** Adolf Friedrich Johann Butenandt (1903–1995) (© akg-images/Fritz Eschen/ picture alliance), **b** silkworm *Bombyx mori* (© irem01/Getty Images/iStock), **c** bombykol

methods, which make it possible to determine molecular structure with just a few milligrams and today even with a few micrograms of pure substance, have become widely used in structural analysis. The final experiment was conducted by Butenandt with more than 500,000 female specimens, from which he removed the last segments of the abdomens and extracted them with a 3:1 mixture of ethanol and ethyl ether. After removing the waxes, he subjected the alcohol fraction to chromatographic purification, and each fraction was tested for biological activity. The active fractions were identified by their ability to induce wing fluttering in males. These were collected, and the scientist isolated 4 mg of a pure substance, which he named bombykol (Fig. 3c) (after the genus name of the silkworm *Bombyx mori*). To the surprise of chemists, bombykol turned out to be a compound with a rather simple chemical structure. Initially, there was a dispute about its spatial configuration, relating to the position of the functional groups. Later, it was established that bombykol is trans-10-cis-12-hexadecadien-1-ol. This was later confirmed by chemical synthesis. Since this formula allows for the existence of four geometric isomers, all four were synthesized. The bioassay showed that only trans-10-cis-12-hexadecadien-1-ol was active. Its activity was comparable to that of natural bombykol. In connection with the elucidation of the spatial structure (isomerism) of bombykol, it was found that this is crucial for biological activity. While the threshold concentration of natural bombykol was 10^{-15} mg/ml, that of the trans-10-trans-12 isomer was 10^{-2} mg/ml (i.e., 10^{13} times lower), that of the cis-10-cis-12 isomer was 10^{-3} mg/ml (10^{12} times lower), and that of the cis-10-trans-12 isomer was 10^{-6} mg/ml (10^{9} times lower).

Let us reflect a little on the threshold concentration of bombykol. If we look at the numbers above, we see an extremely high physiological activity. A concentration of 10^{-15} mg/ml means that 2,500 bombykol molecules are present in 1 cm^3 of air. Taking into account the number of receptors on the antennae of male moths (see below), it can be concluded that only one bombykol molecule per receptor is sufficient to trigger a biological reaction. Other calculations are even more sensational. Considering the amount of substance secreted by a female and the fact that it attracts males from 11 km away, it turns out that bombykol exerts its effect at a concentration of just one molecule per 1 m^3 of air (calculated based on radial diffusion in a hemisphere with a radius of 11 km). Of course, this is not plausible, since the atmosphere is never completely still, allowing bombykol to diffuse evenly in all directions. In the presence of an air current, it will spread more in a cone than a sphere, and its concentration in the airstream will be higher. Once the first molecules of the attractant reach the male's antennae, it flies in the direction of the concentration gradient and thus unerringly finds the female.

Bombykol is the first representative of a large group of physiologically active compounds known as sex pheromones.

In addition to sex pheromones, the group of pheromones also includes other volatile substances of biogenic origin that serve different purposes. These include trail pheromones, aggression pheromones, alarm pheromones, territorial marking pheromones, etc. They are of great importance for animal behavior, and we will address them in the following chapters. Let us now return to sex pheromones.

The discovery of bombykol opened a new chapter in the study of chemical communication between animals and helped clarify the significance of chemical signals for their behavior. Due to their strong physiological effects and low effective threshold concentrations, sex pheromones are synthesized and emitted in extremely low concentrations, making it difficult to isolate them in sufficient quantities for structural studies. For example, 500,000 females of the moth *Lymantria dispar*, 566,000 of *Ephestia kuehniella,* 670,000 of *Pectinophora gossypiella,* 1,200,000 of *Cadra cautella*, etc., were needed to elucidate the chemical structure of their sex pheromones. Only rarely were whole insects used for studies. To improve the purity of the active ingredient, pheromone glands were isolated. However, the purest pheromones are obtained in a completely different way. This is more ethical, as the animals are not killed but placed in an air-flushed chamber. The volatile substances are collected by absorption in a lyophilic solvent (petroleum ether, benzene, etc.) or by condensation at low temperature. The enormous effort does not end with the collection of biological material. This is followed by a long and arduous process of fractionation and purification of the complex natural mixtures, usually by chromatographic techniques. Unlike the routine separation of complex natural product mixtures, the experimenter must keep an eye on the biological activity of the material under investigation at every step, as only in this way can the desired pheromone be identified. The biological test is the only criterion for the presence of the pheromone. In terms of sensitivity, it is surpassed by no chemical or physical method. While gas chromatograph detectors can detect microgram quantities of a substance, the sensitive sensory organs of animals can detect billionths of a microgram of a pheromone. Every chemist is familiar with thin-layer chromatography and knows that it is a very convenient and rapid method for separating organic compounds. Its principle is simple and easy to use. For this purpose, thin layers of adsorbents such as silica gel, aluminum oxide, diatomaceous earth, starch, etc., are used and fixed on a hard, smooth surface (glass or lime). A solution of the mixture to be separated is applied to one end of the prepared chromatography plate using a glass capillary. The plate is carefully placed with the loaded end into a suitable organic solvent (the so-called mobile phase) in a tightly closed container (chromatographic bath), whereupon the solvent begins to rise through the adsorbent by capillary action. On its way upward, it carries the applied substances with it, which move at different speeds depending on their affinity for the adsorbent and their solubility in the mobile phase. When the solvent reaches the other end of the chromatography plate, the components of the mixture are already separated and visible on the plate as chromatographic spots (Fig. 4).

Colorless spots are invisible to the eye but can be made visible with suitable chemical reagents or by irradiation with UV light (Fig. 4). Depending on the method, a few milligrams to several micrograms of a substance can be detected by thin-layer chromatography. When working with pheromones, the resolution and sensitivity of the detection method are of great importance, as their content in crude extracts is extremely low. To increase the sensitivity of the method, especially when searching for female sex pheromones, researchers used the following trick: After drying the plate, they let males

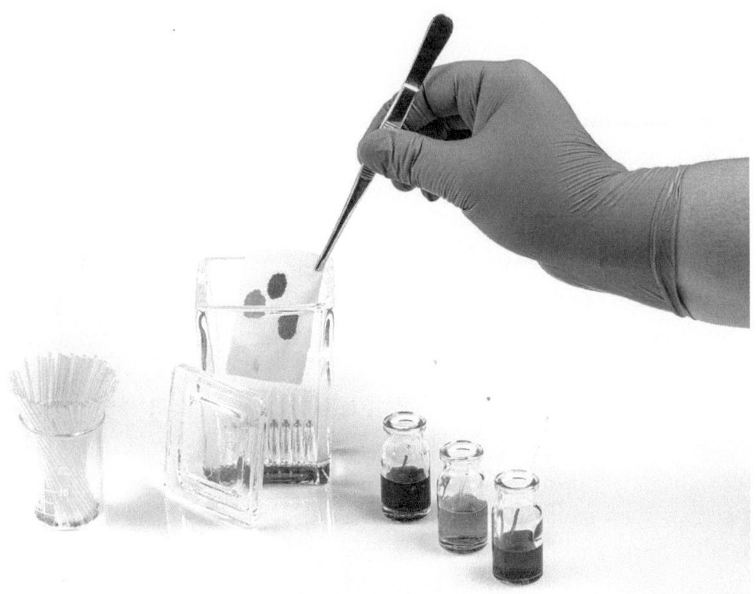

Fig. 4 Thin-layer chromatography: Chromatography plate with applied mixtures of organic substances: After placing the plate in the chromatography chamber, the solvent (the so-called mobile phase) begins to migrate and separates the organic compounds along the plate. (© mehmet/Stock.adobe.com)

walk over it. These unmistakably recognized the spot containing the pheromone and began to dance on it. The experimenter marks the invisible chromatographic spot, scrapes it off together with the adsorbent, and extracts the pheromone with a suitable solvent.

Since it is difficult to obtain highly purified pheromones, their chemical structure is determined using highly sensitive and sophisticated physical methods such as mass spectrometry, nuclear magnetic resonance, etc.

Of all sex pheromones, those of insects, more precisely of butterflies, are the best studied. Chemically, they are unsaturated primary alcohols or carboxylic acids with an unbranched carbon chain consisting of 12–16 carbon atoms. For all pheromones, cis-trans isomerism is important for their biological activity. The exocrine glands that secrete sex pheromones in butterflies are located in the membrane fold between the 8th and 9th abdominal segments. At rest, the gland is hidden. During secretion, the abdomen arches and the gland protrudes outward. The release of the pheromone is facilitated by the numerous folds and tufts on the outer wall of the gland. In some insects, the glands open into the anal opening, while others have no specialized organs for the production and release of sex pheromones. In these cases, certain metabolic products take on the role of attractants. In some beetle species that feed on conifers, for example, the role of sex pheromones is played by terpene compounds similar to those in pine resin. In this case, the formation of pheromones depends directly on the type of food consumed. In the North American moth *Anthera polyphemus*, the formation of sex pheromones requires a specific metabolic factor found in the red leaves of the oak *Quercus rubra*. This “oak factor” was identified as trans-2-hexenal. It is a volatile aldehyde, under whose influence the genitalia of female moths swell and produce sex pheromones.

Sometimes sex pheromones are quite simple substances. In 1970, R. Henzel isolated the pheromone of the New Zealand scarab beetle species *Costelytra zealandica*. He demonstrated that it is the well-known phenol (carbolic acid). Among other things, it is commonly used in households as a disinfectant (Fig. 5). Two years later, the New Zealand scientist D. Hoyt and his colleagues attempted to identify the organ that produces this pheromone. To this end, they isolated a glandular tissue located in the posterior part of the female’s body and found it full of bacteria. Analysis revealed that it was indeed a bacterial colony living in symbiosis with the beetle. After a 5-day incubation in vitro, the bacterial culture began to attract males. This means that the bacteria produce sex pheromones. Analysis of the nutrient medium showed

Fig. 5 *Costelytra zealandica* and the sex pheromone phenol

unequivocally that it contained phenol and other phenolic compounds. According to the authors, the phenol is formed from the amino acid tyrosine, which the bacteria obtain from the proteins on which the beetle feeds.

So far, we have seen that female insects possess sufficiently effective chemical means to signal their sexual maturity and readiness to mate. But what about the males? Do they remain chemically silent during courtship? No. Like the females, males also secrete sex pheromones. In contrast to the female signals, male sex pheromones are less well studied. One of the main differences in the effect of male and female sex pheromones is that the former act over much shorter distances, usually on the order of a few centimeters. When the female perceives the scent of the male, she enters an excited state necessary for copulation. Externally, this state is mainly manifested by the swelling of the abdomen and the extrusion of an ovipositor. After the male has followed the "call of love" and reaches the female, it also triggers her instinct for self-preservation. To avoid injury, she wants to flee at the sight of an approaching body. The task of the male sex pheromones is to suppress this instinct and persuade the female to stay. This makes the sexual act possible in the first place. The male sex pheromones not only serve to induce the female to mate, but also have an attractive effect. Under their influence, the female approaches the male and begins to touch him with her antennae. This increases sexual arousal, which is expressed in a ritual mating dance. Most often, this consists of the male circling the female with fluttering wings. This also has a biological function. By circling the female and fluttering his wings, the male tries to waft as much sex pheromone as possible toward the female, increasing her arousal. In some cases, it has been observed that females lick the pheromone glands of the males, and it even appears that they feed on their secretions. It is believed that, in addition to typical sex pheromones, males also secrete substances with contact effects that further increase the arousal of the females and induce physiological changes that promote the course of copulation.

In 1967, K. Butler proposed that all substances secreted by both sexes that enable mating partners to copulate should be called aphrodisiacs. Sex pheromones can act either as attractants or as both attractants and aphrodisiacs. Conversely, aphrodisiacs can be sex pheromones, but do not have to be.

Very interesting studies have been conducted with male specimens of the genus *Danaus*. A butterfly from the family Nymphalidae, *Danaus gilippus* (Fig. 6), was intensively studied by a team of scientists from the USA and Germany, which included specialists from various fields (chemistry, biology, electrophysiology). It is one of the rare cases in which not only the chemical nature of the pheromones but also their physiological effect could be elucidated. The males of this species have a pair of brushes at the end of their abdomen, similar to those used for coloring eyelashes.

These are connected to the pheromone gland. Two substances have been isolated from the brushes: a ketone derived from pyrrolidine and a divalent unsaturated alcohol related to farnesol. According to electrophysiological studies, only the pyrrolidinone has an attractive effect on the female. However, its significance is not limited to attracting females. It suppresses

Fig. 6 *Danaus gilippus* and its sex attractant pyrrolidinone. (Photo: © leekris/Getty Images/iStock)

their movement reflex and forces them to settle and wait for the male to approach. Since butterflies of the family Nymphalidae (*Danaus*) feed on alkaloid-containing plants, it is assumed that the starting material for pyrrolidinone biosynthesis is the breakdown products of the consumed alkaloids. With the help of chemical compounds, some male insects can also ensure "marital fidelity." The male fruit fly, *Drosophila funebris*, for example, injects, along with the sperm, a polypeptide consisting of 27 amino acid residues, which stimulates the female to undergo renewed fertilization.

Some insects secrete substances during the breeding season that attract individuals of both sexes. These are so-called aggregation pheromones. Their biological significance is that they help concentrate a large number of insects in one place. This greatly increases the likelihood of finding a mate. The role of aggregation pheromones is often fulfilled by metabolic waste products, although some are also known to be produced by specialized pheromone glands. Aggregation pheromones are sometimes also a mixture of several substances that reinforce each other.

This is the case with the bark beetle *Ips confusus* (Fig. 7), which feeds on the bark and wood of the yellow pine. It ejects a huge amount of spent wood (wood dust) along with three terpene alcohols that act as aggregation pheromones. However, all three must be present simultaneously to be effective.

Fig. 7 Bark beetle *Ips confusus* (Sarah McCaffrey, Museum Victoria, CC BY 3.0 au, Wikimedia Commons)

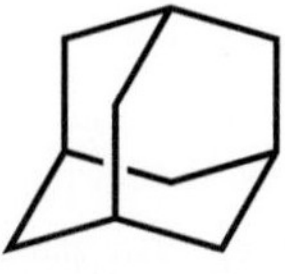

Fig. 8 Adamantane

Bark beetles use the compound adamantane as an attractant (Fig. 8). It was originally found in petroleum and later synthesized. Adamantane smells like camphor, and its molecule resembles the crystal lattice of a diamond, hence its chemical name (from ancient Greek Ἀδάμας, adámas, unconquerable).

How Do Insects Perceive Love Messages?

It has long been known that the sense of smell is important for mating in insects. Male butterflies with amputated olfactory organs were unable to locate the female, whereas those lacking vision mated normally. Where are the olfactory organs of insects located, and how are they structured?

A. Lefèvre demonstrated in 1838 that the olfactory organ of insects is a pair of antennae, located on the top of the head.

The first detailed description of the anatomy of the antennae was provided by G. Hauser in 1880.

The antennae of insects are arranged in various ways. They can be branched or unbranched, may consist of several segments, or taper to a point (Fig. 1). Unbranched antennae generally have a shorter range, like the antennae of female moths, which only detect the males' aphrodisiacs when they are nearby. Conversely, the antennae of males are highly branched, as they must detect low concentrations of pheromones from a great distance. The antennae, like the entire body of the insect, are covered with a hard chitinous shell. Under high magnification, small spine-like protrusions or oval depressions, known as sensilla, can be seen on the antennae. Depending on their shape, sensilla are classified as trichoid, basiconic, igliconic, placoid, etc. Basiconic sensilla are 8–40 µm long (one micrometer equals 10^{-6} m) and are surrounded by a conical chitin cuticle (hard skin), which is part of the overall chitinous covering of the antennae. Under the electron microscope, sensilla are seen to be perforated by numerous pores with diameters between 5 and 100 nm (one nanometer equals 10^{-9} m). Sometimes there are several hundred to several tens of thousands of nanometers. The pores are

I. G. Ivanov, *The Invisible Language of Nature*,
https://doi.org/10.1007/978-3-662-73302-8_4

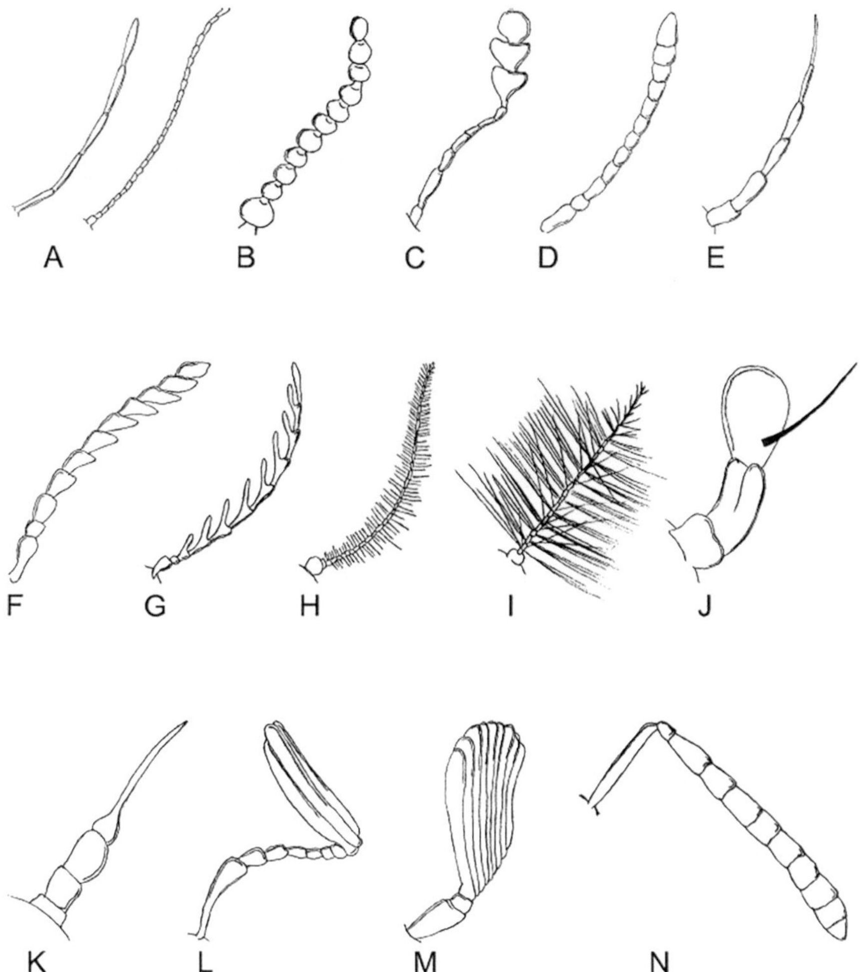

Fig. 1 Types of antennae in insects A = filiform, B = moniliform, C = clavate, D = capitate, E = setaceous, F = serrate, G = pectinate, H = brush-like, I = plumose, J = with arista, K = stylate, L = lamellate, M = fan-shaped, N = geniculate

actually channels through which the molecules of volatile substances reach the receptors. Each sensillum contains one to several dozen nerve cells (neurons) that respond to chemical signals. These are chemoreceptors. Under the influence of chemical stimuli, there is a change in the electrical potential of the receptor, which is referred to as depolarization of the cell membrane and leads to the generation of a signal. The signal is transmitted via a neuron to the nerve ganglion, which in insects serves the function of the brain.

For a long time, the function of the insect olfactory system was studied only through behavioral responses. In this way, the threshold concentration for the effect of numerous chemical compounds was determined. Detailed investigation of the mechanisms by which nerve impulses are generated and transmitted, and analysis of the fine structure of olfactory receptors, only became possible with the introduction of microelectrophysiological methods. With modern electrophysiological techniques, it is possible to measure the electrical potentials of even a single neuron. For this, a silver microelectrode with a tip diameter of 5 μm is inserted into the neuron, and a second electrode is inserted into the nearest blood vessel. The current flowing between the two electrodes is amplified and fed into a recorder, which records the change in electrical potential. The graphical representation of this change is called an electroantennogram (EAG). The amplitude of the EAG is proportional to the concentration of the chemical stimulus. Studies using EAG allow for precise determination of the sensitivity of insects to various chemicals and their threshold concentrations.

Insects live in a world of odors. In addition to their conspecifics, they also detect food sources by smell. It is therefore not surprising that insects can distinguish a large number of volatile substances. The threshold concentration for most of these substances is comparable to or significantly lower than that for mammals and humans.

As already mentioned, insects are particularly sensitive to sex pheromones. Recall that the threshold concentration for bombykol in the silkworm moth is 10^{-15} mg/ml or 4.2×10^{-17} mol/l. In an electrophysiological study, the antennae of *Bombyx mori* were sprayed with bombykol at a concentration of 10,000 molecules in 1 ml. It was found that just one pheromone molecule per antenna is sufficient to generate a nerve impulse. However, at least 200 neurons had to be excited for a reflex response (wing fluttering) to occur. The most potent sex pheromone is that of the cockroach, with a threshold concentration of 10^{-17} mg/ml, i.e., 4×10^{-19} mol/l.

Pheromones are detected by highly specific receptors in the sensilla that are sensitive only to them. In addition to the highly specific pheromone receptors, the sensilla also contain other receptors with lower specificity that respond to other odors. For example, in the sensilla of the pine weevil *Hylobius abietis*, a receptor was found that is excited only by the scent of anethole (the odor-determining substance of anise and fennel). The antennae of the migratory locust *Locusta migratoria* contain four specialized receptors for fatty acids, amines, humid and dry air. The sensilla of the cockroach *Periplaneta americana* also contain four specialized receptors (for pentanol, octanol, butyric acid, and formic acid), and those of the honeybee have nine types of such receptors.

However, the sense of smell in insects differs fundamentally from that of vertebrates (see below). In vertebrates, aromatic substances enter the olfactory organ with inhaled air and flow over the olfactory mucosa in one direction, whereas in insects, molecules of volatile substances fall onto the antennae from all directions. To locate the source of a smell, a vertebrate turns its head and searches for the direction from which the odor comes. This is not necessary for insects. Thanks to the specific structure of their antennae, insects are able to unambiguously localize the odor. They find the correct direction of the odor source without moving their head.

Karl von Frisch, Nobel laureate in Physiology or Medicine in 1973, hypothesized that insects gain a volumetric impression of odor-intensive objects through their sense of smell. He compared their sense of smell to a visual analyzer. Just as a pair of eyes allows us to see the world volumetrically (three-dimensionally), insects possess a pair of antennae that enables them to take a volumetric view of odor-emitting objects. It is hard for us to imagine that an apple "smells spherical" and a cube cut from it "smells cubic," but for insects this is probably possible. This also explains the excellent orientation of bees and ants in complete darkness.

Since When Have Pheromones Existed?

One of the most renowned pheromone specialists, Edward O. Wilson of Harvard University, believes that "pheromones are directly related to hormones." The latter evolved long after pheromones as their analogue for chemical communication between cells in a multicellular organism. Aggregation pheromones are probably the oldest. Even in the era when there were no multicellular organisms on Earth, they helped facilitate communication between cells of the same species. Thanks to them, bacterial conjugation is also possible. Aggregation pheromones also underlie the formation of unicellular colonies, some of which are so stable that they are also referred to as multicellular organisms. These include, for example, the flagellates of the genus *Volvox* (Fig. 1).

Their colonies are often found in freshwater swamps and lakes. They look like small spheres about 1 mm in diameter, from which numerous microscopically small flagella protrude. Observation under the microscope shows that the colony consists of about 1,000 cells embedded in a semi-liquid, jelly-like mass. The cells are connected to each other by cytoplasmic bridges, allowing synchronization of their activities. During reproduction, some cells sink into the interior of the sphere. There, they begin to divide and form smaller colonies. These then separate from the mother colony. Maintaining a multicellular colony is impossible without the involvement of aggregation pheromones.

Aggregation pheromones also underlie the life cycle of another interesting species, the amoeba *Dictyostelium discoideum* (Fig. 2). It represents the rare case in which an organism can exist in two forms—unicellular and

I. G. Ivanov, *The Invisible Language of Nature*,
https://doi.org/10.1007/978-3-662-73302-8_5

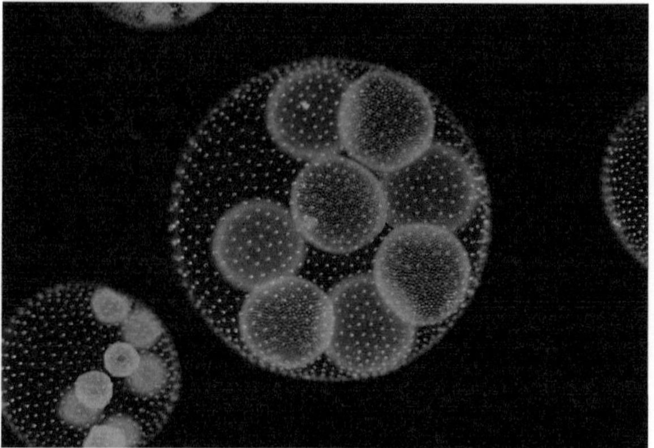

Fig. 1 *Volvox*. (© micro_photo/Getty Images/iStock)

multicellular. In its unicellular form, it resembles an amoeba—it reproduces by division and moves by protrusions of the cell wall (so-called pseudopodia). When the environment becomes depleted of nutrients, some cells begin to secrete pheromones. Other cells are attracted, and a colony forms that resembles a plasmodium. In parallel with aggregation, cell differentiation occurs. Some cells form a stalk, others a fruiting body. From the cells at the tip that form the fruiting body, true spores with a cellulose shell develop.

Pheromones can also determine the sex of some lower eukaryotes. The hyphae (filaments) of the fungus *Achlya bisexualis*, for example, are asexual but are referred to as male or female depending on the sex of the nearest hypha. Thus, if an undifferentiated hypha happens to be adjacent to a female hypha, branches form on it, which transform into antheridia (male sex organs) and produce spermatozoa. At the same time, oviducts containing eggs form on the female hyphae. The trigger for this complex process of sexual differentiation is a pheromone from the class of steroids (antheridiol) secreted by the female hyphae (Fig. 3). The male hyphae, in turn, secrete another pheromone that stimulates the formation of oviducts. A similar phenomenon is also observed in *Volvox aureus* colonies.

Exactly when life originated on Earth is uncertain. The first heterosexual organisms developed. Reproduction occurred through the formation of two types of sex cells (gametes), from which, after fusion (fertilization), a new organism emerged. How do these microscopically small cells find each other in an immense watery environment?

It is likely that, at this crucial moment for life on the planet, sex pheromones arose in nature, i.e., specific chemical signals secreted by the egg cell

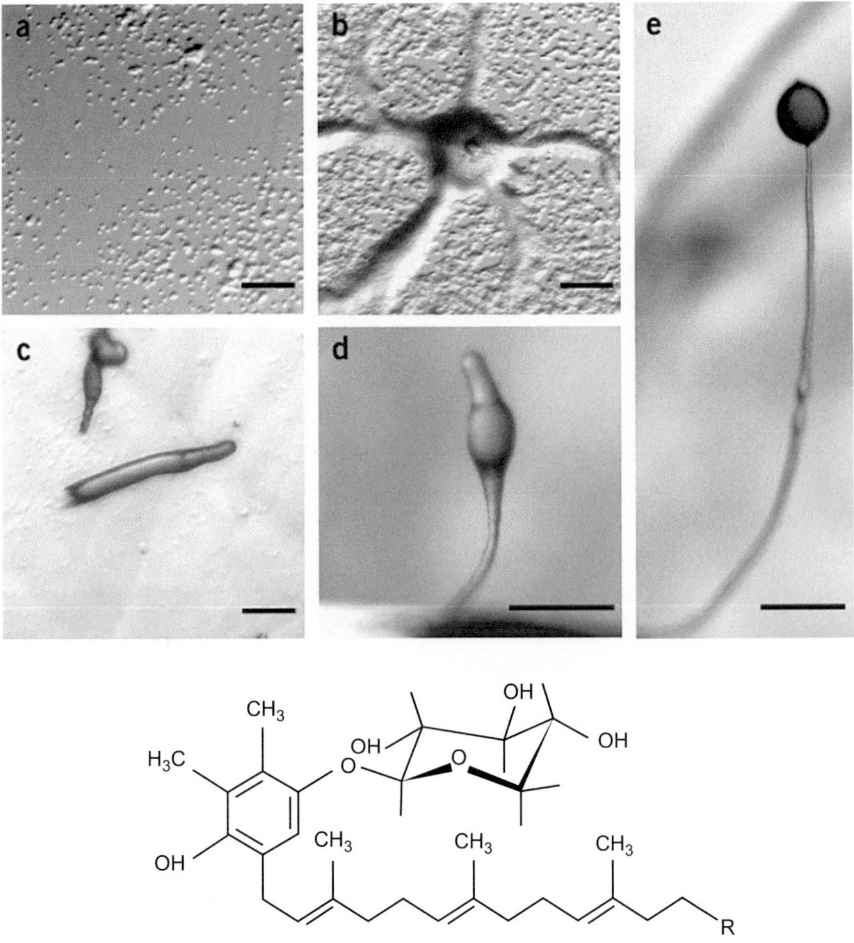

Fig. 2 *Dictyostelium discoideum* and aggregation pheromone

Fig. 3 Sexual pheromone antheridiol

that stimulate the male sex cell to move against concentration gradients and reliably recognize the egg cell. In this case, the female germ cell does not need to move or actively search for the male. On the contrary, if it remains stationary, the likelihood of being found increases. This is why female gametes lack motile flagella, which are an obligatory feature of sperm cells.

For the first time in 1968, a chemical substance was isolated from the subtropical fungus *Allomyces macrogynus* (Fig. 4) that is secreted by female gametes and attracts male sex cells. The fungus grows on the carcasses of insects and other animals and forms hyphae up to 1 cm long. Mature fungi release male and female germ cells. The males are orange due to their β-carotene content and are 2 to 3 times smaller than the females. From female gametes, the substance sirenin was isolated (Fig. 5), which is a strong attractant for male gametes. It is a divalent sesquiterpene alcohol.

The attractiveness of sirenin is evident in that even concentrations of 10^{-10} mol/l, i.e., about 0.02 mcg/ml, are still effective. A comparable substance was also found in the brown alga *Ectocarpus siliculosus*. It has long been observed that the egg cells of this alga strongly attract male gametes. These accumulate in the algae and begin to touch them with their flagella. After fertilization, the resulting zygote (fusion of the nuclei of the male and

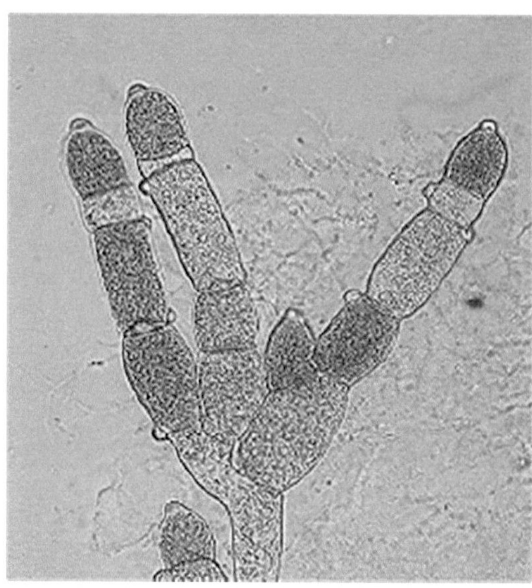

Fig. 4 *Allomyces macrogynus*

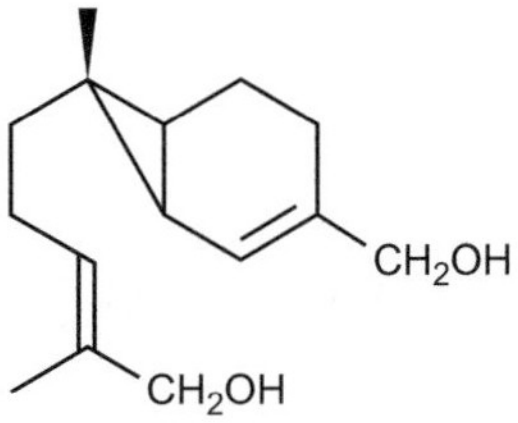

Fig. 5 Sirenin

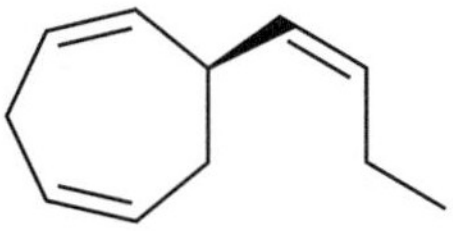

Fig. 6 Ectocarpen

female germ cells) no longer has any attractive effect, and the male gametes quickly disperse.

To identify the attractant, biomass obtained from 14,900 Petri dishes was used. From this, 92 mg of a substance called ectocarpen could be isolated. It is an unsaturated hydrocarbon with 11 carbon atoms and a 7-membered ring (Fig. 6). Pheromones of female sex cells have also been isolated from many other lower organisms, especially algae and fungi. They are probably also prototypes of sex pheromones in higher organisms, and it is likely that similar substances are involved in fertilization processes in higher organisms. How else could the relentless drive of sperm toward the egg and their consistent movement against the flow of physiological fluids with the sole aim of fertilizing it be explained? It is unlikely that, in addition to the chaotic movements of sperm and a chance encounter with the egg, no other mechanisms would have evolved.

Why Do We Study Sex Pheromones?

When Adolf Butenandt reported in one of his lectures about his long and intensive work in the search for and study of bombykol and wrote its relatively simple chemical formula on the board, a female student asked what was so interesting about it. After all, bombykol was just a chemical compound.

To answer the question, let us briefly consider the significance of pheromones for humans, with a focus on insect pheromones. If the student regarded bombykol as merely a chemical compound, she was correct in that respect. Pheromones are indeed attractants for insects, but not for chemists, since their formulas are often simple. Even when a researcher encounters a new compound unknown to science, with a beautiful and exotic formula that brings professional satisfaction, pheromones are a challenging subject of study. They are primarily of interest to scientists who are more concerned with their effects than with their chemistry.

With the discovery of pheromones, many puzzling phenomena of classical biology found their explanation. They have proven to be a powerful tool with previously unimagined capabilities for maintaining biological balance in ecosystems. This makes pheromones a discovery of epochal significance for ecology, ethology (behavioral research), and animal physiology. In addition to fundamental information, the study of pheromones is also of practical importance. Sex pheromones can be used to control pests of crops and insects harmful to humans. There are two possibilities for their practical application: a) as bait to collect insects and kill them with conventional insecticides, b) by dispersing them in high concentrations in the air to

I. G. Ivanov, *The Invisible Language of Nature*,
https://doi.org/10.1007/978-3-662-73302-8_6

suppress the natural signals of mates during reproduction. The possibility of using female insects as bait to kill males was first employed in 1930–1931 by the Czech entomologist (insect specialist) P. Dick. He placed cocoons of females of the moth species *Lymantria monacha* (nun moth, Fig. 1) in boxes, hung these on trees, and attached adhesive strips nearby. After the females emerged, the males gathered around them and became stuck to the strips. This method became known as "Dick's Method of Monk Butterfly Control."

In 1940, V. Ambros applied Dick's method to clear a 756-hectare area infested by a moth, using 480 matchbox traps. In this way, he was able to destroy 384,448 males within 49 days. To prolong the effectiveness of the bait, he placed cotton in the boxes, which continued to attract males for 1 or 2 weeks after the death of the females. Later, Dick's method was also used to control other insects. Over time, it was modified by replacing the live females with synthetic pheromones (Fig. 2). This made it possible to extend the attractiveness of the traps indefinitely. In addition, the adhesive strips were replaced by highly effective insecticides, further increasing the method's efficacy. By using aggregation pheromones as bait, not only males but also females can be attracted and destroyed.

The second approach to using pest pheromones is the massive deployment of female attractants. We have already seen that the effective threshold of female sex pheromones is extremely low. When a pheromone is released into the atmosphere at a concentration a hundred or a thousand times above its threshold, the males become confused and cannot locate the females. Even if they were to find them, in their overstimulated state they would be unable to successfully copulate.

Fig. 1 Nun moth *Lymantria monacha.* (© neil bowman/Getty Images/iStock)

Fig. 2 Pheromone trap. (© Ivan Lopez Gonzalez/Getty Images/iStock)

Advantages of Pheromones over Insecticides

In the past, the control of insect pests was carried out using chlorinated hydrocarbons such as DDT, hexachloran, etc. These are relatively non-toxic to warm-blooded animals. However, because they are chemically stable, this has led to their accumulation in large quantities in soil and water, and consequently they are also found in animal-derived foods. As a result, their production and use have been banned worldwide. They have been replaced by organophosphorus insecticides, which are less stable and break down more quickly in natural moisture. Compared to chlorinated hydrocarbons, they are significantly less toxic to warm-blooded animals and humans. Both groups of insecticides have a broad spectrum of activity and kill both beneficial and harmful insects. They often do more harm than good, especially when the ecological relationships between organisms are ignored.

Aphids are a scourge worldwide and are combated everywhere. They have natural enemies such as ladybugs, predatory bugs, and wasps, which often manage to keep them in check. However, aphids are naturally resistant to many synthetic insecticides, as these penetrate their bodies less easily than those of the insects that feed on them. Therefore, broad-spectrum insecticides primarily kill aphid predators (entomophages, insectivores) and to a lesser extent the aphids themselves. As a result, the number of aphids has increased sharply in recent years. One possible way out of this situation is the search for insecticides with higher selectivity, such as amiphos. It is 300 to 500 times more toxic to beet aphids than to their natural enemy, the seven-spot ladybird. However, the search for agents with higher selectivity is very expensive. According to the US Bureau of Statistics, it takes about 10

I. G. Ivanov, *The Invisible Language of Nature*,
https://doi.org/10.1007/978-3-662-73302-8_7

years and costs hundreds of millions of dollars to bring a new insecticide into practical use.

In contrast to insecticides, pheromones are species-specific. This makes it possible to control the population of a particular biological species without affecting the number of beneficial insects. As already mentioned, pheromones are relatively simple chemical compounds whose synthesis is no more complicated or expensive than that of insecticides. However, they have the great advantage of being absolutely harmless to other organisms and not polluting the environment. Even when used in combination with insecticides, they have the advantage of selectively attracting and killing only a specific type of insect, and with a much smaller amount of insecticide. The insecticide "Fly-Tox" has long been used in the USA in combination with the housefly's sex pheromone. It is significantly more effective than Fly-Tox without the pheromone. It can be assumed that synthetic sex pheromones will soon find even broader application as baits for insect control, both in domestic and industrial settings.

Chemistry of the Model Society

A society without selfishness, envy, baseness, and corruption, without rich and poor, without privileged and despised, a society of industrious beings living in peace and understanding, where everyone knows their duties and is ready to sacrifice their life for others. This is the ideal society envisioned by idealistic philosophers, religious leaders, writers, and other good people. Thousands have given their lives for this society, without ever experiencing it. In real life, this society of *Homo sapiens* does not exist. In nature, however, it is not a chimera. This is how colonies of social insects such as bees, wasps, ants, and termites live. In their astonishing perfection, humans may rightly envy them. The lives of these insects have always aroused human curiosity. Generations of scholars have entered their mysterious world and marveled at their perfection.

In many respects, the behavior of social insects resembles rational human action. For example, they build and maintain their homes in exemplary fashion and divide their tasks among themselves in the most rational way. There is true specialization and division of labor, as well as precise coordination between the various types of work (construction and repair, food gathering, brood care, security, cleaning, etc.). And since humans tend to seek a reason for any observed order in everything that does not descend into chaos, people long regarded social insects as rational beings. But today, to speak of reason is uninformed. We know that our reason requires a complex nervous system and a highly organized brain. Bees, ants, and termites do not possess this. Their ganglia (the analogue of a brain) do not differ from those of other insects. Among bees, of which there are about 20,000 species, only

I. G. Ivanov, *The Invisible Language of Nature*,
https://doi.org/10.1007/978-3-662-73302-8_8

about 10% are social, and their nervous system is in no way superior to that of non-social bees. If we can speak at all of a connection between "intelligence," nervous system, and way of life, then ants are the most intelligent among social insects. In the heads of most insects, there are two nerve ganglia, an anterior and a posterior. The anterior acts as the brain, and the posterior controls the jaws of the mouth. In ants, the two ganglia are fused into a single organ, apparently because their feeding habits have allowed a significant simplification of the head. But whatever the reasons, ants are among the insects with the largest "brains." In them, the social way of life is also the most highly developed. While in bees and wasps, social organization is an exception (considering the large number of non-social species), in ants it is the only form of existence. According to the observations of some entomologists, ants are the insects that can be "trained" most easily. However, these peculiarities have nothing to do with reason.

All activities inherent to social insects are carried out mechanically, based on a pre-formed, genetically determined program. Jean-Henri Fabre writes: "Instinct is infallible only within its assigned domain. Outside of it, it is powerless." To test the intellect of the wasp, Fabre conducted a simple but illuminating experiment. He observed that one of the predatory wasps, which feeds on grasshoppers, first paralyzes its victim with a sting and then drags it to the nest by the antennae of the head. When he cut off both antennae, the wasp bit the victim by the labial palps and dragged it away again. Then the scientist also cut off the mouth palps. The wasp eyed the mutilated victim and, without hesitation, left it behind, even though there were still many protrusions on the body that could have been bitten and used for dragging. But, as Fabre writes: "To grasp a leg instead of an antenna presented her with an insurmountable difficulty. All she needed were antennae and palps. If these were to disappear from the heads of her victims, the wasp lineage would also disappear, unable to overcome the slightest difficulty."

Thanks to instinct, every social insect can flawlessly fulfill its tasks, but instinct alone is not enough to secure the life of the society as a whole. All types of activities must be harmonized to ensure existence. A mechanism is needed to switch blind instincts on and off as needed and to synchronize the activities of all family members. Something is needed to unite the insects and give them the feeling of being part of a collective, equivalent to an organism. For a long time, this mysterious force, capable of uniting thousands of irrational creatures into a living and rational collective whose overall activity is comparable in complexity to that of humans, remained a mystery to science.

In describing the life of bees, Maurice Maeterlinck spoke in 1901 of the "elusive social forces" that govern the activities of the bee colony. He wrote: "Where is the spirit of the hive, which disposes of the wealth and happiness, the freedom and life of every winged being within it?"

It later became clear that this mysterious "spirit" is a complex of chemical compounds secreted by the eccrine (externally secreting) glands of social insects. In fact, these are the pheromones already known to us. To illustrate their significance for the life of insect communities, let us take the example of honeybee colonies, which are the best studied. Here we run the risk of boring those readers who are familiar with the structure and organization of the bee colony. However, we believe this is necessary for a better understanding of the importance of pheromones for the life of bees.

The honeybee *Apis mellifera* has always been a true domestic animal. It pollinates fruit trees and provides humans with honey, wax, propolis, and bee venom. In the wild, it is now rarely found. A normally developed honeybee colony consists of about 60,000 bees and includes three types of bees: one queen bee (also called the queen), 100–200 male bees (drones), and all the rest—the worker bees. The three bee types can be easily distinguished by their appearance (Fig. 1). The queen has an elongated, slender body, about 1.5 times longer than that of the worker bees. The body of the drones is thicker and rounder.

Although both the queen and worker bees are female, the only fully developed female is the queen. Her main task is to lay eggs. A healthy and young queen lays 1,500–3,000 eggs daily, with a total weight equal to her own. Unlike most other animals, however, she has no maternal instinct. Once the

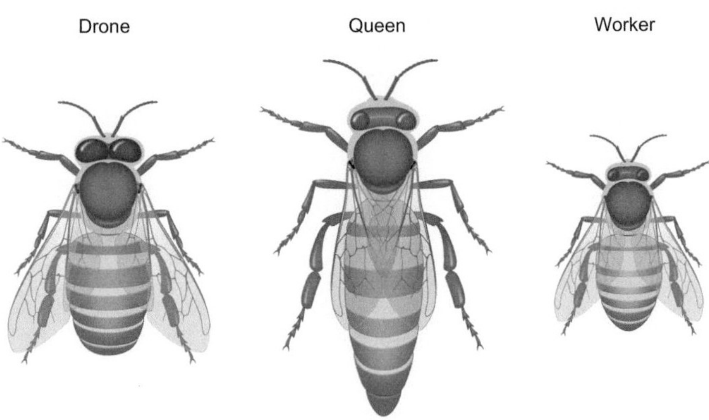

Fig. 1 Types of bees in the honeybee colony *Apis mellifera*. (© Aldona/Getty Images/iStock)

egg is laid, the queen no longer cares for it. Maternal care is provided by the nurse bees, the worker bees. The drones have the least task. Their role is limited to fertilizing the queen, which is performed by one of the drones. Afterward, they are no longer needed and are usually thrown out of the hive by their sisters. The strict law of the bees, "He who does not work, shall not eat," applies to these "blood brothers" without compromise.

All other activities in the hive—building brood combs, preparing cells for egg laying, maintaining hygiene and constant temperature in the hive, collecting nectar and pollen, rearing the brood (larvae), guarding the hive against enemies, etc.—are performed by worker bees. It was once thought that each of these activities was carried out by "special specialists." Today, it is known that every worker bee goes through all units of work. On the third day after the egg is laid, a small larva hatches, which is fed by the nurse bees. This is no easy task. The larva is so voracious that it needs about 1300 meals a day to be satisfied. Thanks to its good appetite and rich diet, it grows quickly and fills its hexagonal home in 6 days. Then it spins a silk cocoon around itself, which is sealed by the worker bees with a thin wax cap, so it can develop into a bee undisturbed. This process takes another 12 days, so the worker bee hatches 21 days after the egg is laid. A queen does so about 5 days earlier, the drone 3 days later. Immediately after leaving the brood cell, the worker bee actively participates in hive life. Her first task is hygiene. She cleans the cells of the larvae and prepares them for the storage of honey or for egg laying. After about 6 days, the worker switches to a new task—"nurse"—and takes care of feeding the larvae. After another week, she becomes a builder and is responsible for repairing old and constructing new wax chambers. This phase of her specialization lasts about 10 days. During this time, she also performs some other tasks: she receives nectar from the forager bees and processes it into honey, stores honey and pollen in the cells, cleans the hive of superfluous objects and dead bees, stands "guard" at the "checkpoint" (the brood chamber), etc. During this period, she is a stinging bee as the guardian of the colony. Only after the 20th day after hatching does the worker bee leave the hive and go foraging. But this is actually the time of her old age, for in the active summer season, the worker bee lives only 5–6 weeks. The autumn bees live until spring.

Apparently, the bee is born with pre-formed, genetically determined behaviors. But how does she know when to leave the hive and take on a new task? Nature decides this in a very rational way. The bee is simply born underdeveloped. Her development continues after leaving the brood cell. On the 6th day, for example, her salivary glands develop, which secrete royal jelly ("queen bee food juice") with which she feeds the larvae. These glands are functionally comparable to the mammary glands of mammals. During this time, the wax glands are still underdeveloped. On the 14th day, the salivary glands have atrophied, but the wax glands develop. This automatically determines the worker bee's new occupation. After the 20th day, the wax glands regress and the bee begins to collect nectar and pollen.

We have already said that both queen and worker bees are female and drones are male. In bees, sex is determined by a haploid-diploid mechanism, which means that fertilized eggs produce females and unfertilized eggs produce males. Whether an egg is fertilized or not is decided by the queen bee. During the mating flight, the fertilizing drones fill her spermatheca with millions of sperm, which remain alive for about 4–5 years. The spermatheca is connected to the ovipositor by a thin, sphincter-equipped duct, and during egg laying, the queen can "at will" release sperm to the egg or not.

Since drone larvae are larger than those of the workers, special cells are built for them. Depending on the size of the cell, the queen circles around it and lays a fertilized or unfertilized egg.

And from which eggs does the queen herself hatch? Strangely enough, she comes from the same eggs as the worker bees. Whether an egg becomes a worker or a queen depends only on the diet and thus on the "will" of the nurse bees. In the first days after hatching, all larvae are fed royal jelly. It is very nutritious and extremely rich in vitamins and other biologically active substances. As the worker bee larvae grow, they are switched to a different diet—royal jelly supplemented with honey and pollen. Only the queen larvae continue to be fed royal jelly until the end. To this is added a still unknown biologically active substance to stimulate the development of the reproductive system. It is estimated that 1 kg of royal jelly contains only 1 mg of this substance. Since the queen larvae are much larger than those of the worker bees, special queen brood cells are needed, which have the shape and size of an acorn. These are usually located at the bottom of the comb. The crowning of the queen is thus entirely in the hands of the worker bees. However, they do not often replace their queens. The preparation for a queen change, recognizable by the construction of new brood cells, is an alarm signal and an indication that something is wrong in the hive. Like all other activities of the bees, the rearing of queen larvae is nothing but instinct. But why does this instinct not always appear, but only at certain times? What mechanism switches it on and off, and who determines exactly when to start rearing a new queen? The answer to this question was provided by two laboratories. One in England, the other in France.

Independently of each other, they succeeded in isolating an extremely important chemical compound from the mandibular glands (salivary glands) of queen bees. It was identified as 9-keto-2-trans-decenoic acid = 9-oxo-trans-2-decenoic acid (Fig. 2) and called queen substance.

Research shows that the queen substance is the magnet that keeps the worker bees near the queen. It is the magic wand with which the queen bee brings order and peace to her colony. The queen substance is a pheromone, but unlike other typical pheromones, it is characterized by a broader spectrum of action. It also differs from pheromones in the way it is distributed. A portion of the substance is distributed in the hive through the air and, like sex pheromones, is perceived via the antennae. It has a strong attractive effect and causes the bees to cluster around the queen for care. Through this contact and the constant contact among themselves, during which water, nectar, and pollen are exchanged mouth-to-mouth, they spread the queen substance among themselves. This phenomenon is widespread among social insects and is called trophallaxis (transfer of liquid food). This type of transmission of the queen substance was initially hypothetical, but has since been experimentally demonstrated using the method of labeled molecules. For this purpose, several bees were given queen substance labeled with radioactive isotopes. This could be detected in almost all bees in the hive after just a few hours. Radioactive isotopes also showed that the queen substance is metabolized relatively quickly. Within a few hours, it is converted into 9-hydroxydecanoic acid and 9-hydroxydecenoic acid. The rapid metabolism, in turn, requires that the substance be produced continuously and in sufficiently large quantities. It was found that for the normal operation of an average-sized hive, about 0.1 mg of queen substance per day is required.

The queen substance can also be compared to a sedative. It calms the bees' nervous system and creates the conditions for normal work. But it also resembles a drug, to which they become strongly dependent. Without the queen substance, the bees experience withdrawal symptoms, become restless, agitated, wander aimlessly around the hive, buzz, and flutter their wings. This picture is always observed when the queen is removed. Due to the rapid metabolism of the substance, the entire hive knows 2–3 hours after the queen's removal that she is gone. The substance also has the ability to suppress a number of bee instincts, above all the instinct to form queens and feed queen larvae. With the removal of the queen, these instincts are

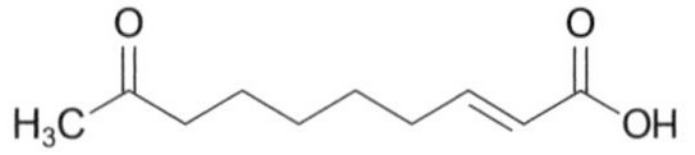

$$H_3C-C(=O)-(CH_2)_5-CH=CH-C(=O)OH$$

Fig. 2 The queen substance of *Apis mellifera* (9-keto-2-trans-decenoic acid)

released, and the hive begins to prepare the crowning of a new queen. How does this happen?

The bees hurriedly build several brood cells, into which they transfer one of the most recently laid eggs (no older than 3 days) and begin to feed the freshly hatched larvae with royal jelly, mixed with some of the aforementioned unknown active ingredient. Thus, after just 16 days, they already have a new queen. To avoid the risk of failure in queen production, the bees create several brood cells at the same time, from which several queens temporarily hatch. However, there is only one throne. Therefore, the first queen bee, after hatching from the brood cell, devours the other brood and thus kills her rivals. If one of them survives, a fierce fight ensues, ending with the death of one of them.

On the sixth day after hatching, the young queen is mature and embarks on a mating flight, including several orientation flights. The signal to depart is given by the sex pheromones. It has been found that the queen substance also acts here. In addition, during this time, another substance with a similar chemical structure (9-hydroxy-2-trans-decenoic acid) and possibly as yet unidentified synergistic substances are secreted. The fertilization of the queen, which lasts only a few seconds, is carried out in the air by one of the accompanying drones. It has been observed that the queen attracts the drones most strongly at a height of about 12 m. Below 4.5 m and above 30 m, they no longer follow her.

The queen substance also acts like an antihormone, suppressing the functions of the worker bees' reproductive system. As already mentioned, they are female and have the potential to lay eggs. In the presence of a queen, this does not happen. The queen substance completely suppresses this activity.

As mentioned, the bees immediately produce a new queen if the hive is left without a queen (due to death or experimental removal). But what if there is no suitable, freshly laid egg available for this process?

After being freed from the suppressive effect of the queen substance, the reproductive organs of one or more worker bees rapidly develop and they begin to lay eggs. However, their eggs are unfertilized and from them hatch drones, which quickly consume the collected honey. This usually leads to the death of the hive.

The amount of queen substance in the hive is an indicator of the queen bee's vitality and her suitability to be queen. In the case of illness or old age, the production of queen substance decreases, and the hive becomes less active. This is a sign that the bees need to produce a new queen.

Some authors assume that the queen substance is also the trigger mechanism for bee swarming. What is swarming? In the months of May or June, when the number of bees in a colony with a healthy queen increases too much, the bees begin to lay new eggs to produce a new queen. The two queens do not come into conflict, but at a certain point, at a specific signal, the old queen leaves the hive together with part of the bees (20,000–30,000) and makes room for the new queen. The departing swarm floats through the air like a flying ball and settles on a tree, a roof ledge, a bush, an abandoned hive, a wooden box, etc. Who decides that the hive is overcrowded?

It is assumed that when the number of bees exceeds a certain limit (80,000–90,000), the old queen cannot provide enough queen substance to meet the demand. The worker bees and those further from the nest develop the feeling of being queenless.

Studies on the structure-function relationship of the queen substance show that its chemical structure is unique to its biological activity. Attempts to lengthen or shorten the carbon chain, to replace the trans-isomer with a cis-isomer, to change the position of the carbonyl group, etc., have led to a loss of activity.

In different social insects, the chemical nature of the queen substance varies. In the wasp *Vespa orientalis* (Oriental hornet), for example, it is a lactone of 5-hydroxyhexadecanoic acid.

Pheromones are secreted not only by the queen but also by the worker bees. Each of them carries a "perfume bottle" at the end of their abdomen. This is a small organ called the Nasonov gland, named after the Russian zoologist N.V. Nasonov (also Nassanoff). It is located on the back of the last abdominal segment and can be recognized in bees as a light spot (Fig. 3). The gland secretes aromatic substances that are of great importance for the life of the bees. It is their identification card, with which the worker bees are

Fig. 3 The Nasonov gland (yellow spot) of *Apis mellifera* and the pheromone geraniol it secretes. (Photo: © Aleksandr Rybalko/Getty Images/iStock)

admitted to the hive. With the help of the Nasonov gland, strangers are easily recognized and expelled. The scent is strictly specific to each bee.

The Nasonov gland is an important organ of the bee colony, with which the worker bees orient themselves after returning from foraging or the queen after the mating flight, to find their home. With the secretions of the Nasonov gland, scout bees also mark newly discovered food sources, which in turn facilitates the work of the nectar and pollen collectors. It is known that scout bees communicate information about the location of the food source to the foragers by means of the so-called "waggle dance." They mark certain colors with the pheromones of the Nasonov gland. The main component of the secretion is geraniol. It is a terpene alcohol found in many essential oils, including that of the rose. In addition, the gland also produces other terpene compounds such as citral, nerolic acid, geranic acid, etc. It is likely that the unique character of a bee colony's scent is determined by the variation in the ratio of these compounds.

In 1814, J. Hübner observed that when a person is stung by a bee, the likelihood of being attacked by a second bee is much greater than if not stung at all. He also noticed that the sting site had a fruity smell reminiscent of banana aroma. The secret of this scent mark was only revealed much later. It was found to be isopentyl acetate. It is the product of another pheromone gland associated with the sting (Fig. 4). Its purpose is to mark enemies and designate them as targets for other bees. With each additional bee that leaves its "scented" autograph with its sting, the concentration of isopentyl acetate increases and fuels the bees' aggressiveness. It has been experimentally shown that filter paper or a piece of fabric soaked with isopentyl acetate is attacked and stung by bees. Perhaps this explains a long-known fact, namely that bees do not like drunken beekeepers. The reason is probably that spirits (wine, schnapps, cognac, etc.) contain, among many other fruity-smelling esters, isopentyl acetate.

Newly hatched bees do not secrete isopentyl acetate. It appears around the 15th day of their life, and the content varies between 1 and 5 μg per bee. It is believed that isopentyl acetate serves only to mark enemies in whose bodies the bee leaves its sting. When other insects are stung, the sting is withdrawn and isopentyl acetate is not released. Small pests are marked with another pheromone, 2-heptanone.

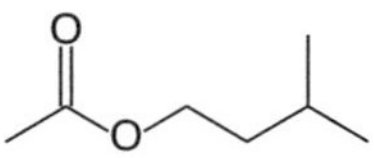

Fig. 4 Isoamyl (isopentyl) acetate, alarm pheromone of *Apis mellifera*

It has long been known that smoke reduces the aggressiveness of bees. That is why a source of smoke is an obligatory attribute of a beekeeper. The effect of smoke is not precisely known, but there are two hypotheses. One suggests that smoke is perceived as an alarm signal (probably in connection with the natural disaster of fire) and the bees instinctively begin to fill their abdomens with honey to prepare for evacuation. A full abdomen, however, makes stinging more difficult. According to the second hypothesis, components of the smoke bind to the isopentyl acetate receptors located on the antennae, making them insensitive to the alarm pheromones.

Although so far we have only considered the role of pheromones in the life of the honeybee, their importance for other social insects is no less great. In these as well (wasps, ants, and termites), the unity and cohesion of the colony are due to pheromones secreted by a queen and distributed among the colony members. In addition to the aggregation function, pheromones also have other physiological effects. In termite and ant communities, which are characterized by a caste society, they also regulate the number of castes.

Through pheromones, non-flying insects mark their path to newly discovered food sources. These are so-called trail pheromones. The American fire ant *Solenopsis saevissima* leaves a scent trail on the way to the anthill after foraging by touching the ground at regular intervals of a few centimeters with its sting. All other ants that follow this trail quickly find the sought-after food source. On the way back, the path is also marked. This strengthens the scent of the trail and attracts new ants to the food depot. When the food source is exhausted and the ants return without food, they no longer leave a trail. The scent dissipates and disappears completely after a few minutes. In this way, it is possible to determine exactly how many workers are needed to transport the food from the source to the anthill, without wasting unnecessary labor.

The scent trail of ants is not just a corridor connecting the anthill with the food source. It also has a vector character, i.e., it has a direction. This was demonstrated as follows. A piece of paper was placed in the path of the ants, which the workers marked with pheromones. If the paper is rotated 180° at a certain moment, some of them go to the end of the sheet and turn back, while others remain on the sheet until they find the continuation of the trail outside the sheet and continue on their way. It is assumed that the concentration of the scent increases in the direction from the anthill to the food source. This is not insignificant, as the strengthening of the scent is a sign that the ants should move faster toward the goal. It is possible that the invisible trail, in addition to direction, has a more complex structure and contains much more information than we can imagine. For example, McGregor observed that ants always pass the same point on their way to the anthill and that those who accidentally get lost wander for a long time until they find the right path.

To fulfill their purpose, trail pheromones must be volatile or evaporate relatively quickly. Extracts of the red wood ant *Formica rufa* have been shown to retain their effect for up to 3 years at 4 °C, whereas at 25 °C they remain active for only a few hours. In leafcutter ants (*Atta texana*), the trail pheromone is the methyl ester of 4-methylpyrrole-2-carboxylic acid (Fig. 5). The termite *Zootermopsis nevadensis*, on the other hand, uses caproic acid for the same purpose, and *Kalotermes flavicollis* (yellow-necked drywood termite) employs cis-hex-3-en-1-ol.

Long before trail pheromones were identified, they were well known to insectivorous animals. Beetles that feed on ants from the families Histeridae (clown beetles) and Staphylinidae (rove beetles), millipedes, and even snakes can easily locate ant nests by following their scent trails.

All social insects possess reliable individual means to defend themselves and fight off enemies. However, their fighting power is much more effective when they join forces. Such cooperation is also necessary in the event of accidental disasters such as fire, flooding, nest destruction, and so on. The mobilization of the community's defensive forces is achieved only through pheromones. These are the so-called alarm pheromones. Their presence in the air signals danger and leads to a sudden change in the work program. The insects abandon their current activities and go into a state of combat readiness.

Fig. 5 The trail pheromone of the leafcutter ant *Atta texana* is the methyl ester of 4-methylpyrrole-2-carboxylic acid. (Photo: © NokHoOkNoi/Getty Images/iStock)

When an ant patrolling the nest or its vicinity notices something disturbing, it releases an alarm pheromone and thus sounds the alarm. The pheromone quickly spreads over a distance of 6–10 cm, putting the ants in this area into an excited state. If these ants also perceive the situation as alarming, they too release alarm pheromone. The concentration in the air rises rapidly, causing the radius of the active zone to expand. In this way, the signal is amplified, and within a few seconds the entire ant nest is alerted. However, if it is a false alarm, the ants nearby do not confirm it, the pheromone scent dissipates, and the signal quickly fades.

Ants possess two alarm pheromone glands: the mandibular gland (in the upper jaw) and the Dufour's gland (in the abdomen, named after the French entomologist Léon Dufour, 1780–1865). The mandibular gland produces terpene compounds such as citronellol, citronellal, norcitronellal, geraniol, and so on, while the Dufour's gland secretes paraffin hydrocarbons and methyl ketones. In the event of an alarm, both glands are activated, producing a complex mixture of several volatile organic compounds. For example, 46 such compounds have been identified in the alarm signal of the red wood ant. Since the signal consists of multiple components, its quantitative composition can vary greatly depending on the reported situation, allowing the insects to convey additional or clarifying information about the danger. They may possibly transmit, along with the alarm signal, information about the type of threat, the extent of damage to the nest, the nature of the enemy, and so on. In different insects, alarm pheromones have different chemical compositions. As already mentioned, the alarm pheromone in the honeybee is isopentyl acetate, while in termites and ants it is terpene compounds. In the Australian termites *Drepanotermes rubriceps* (Fig. 6) and *Drepanotermes perniger*, the alarm pheromone is D-limonene; in *Amitermes meridionalis* it is terpinolene (Fig. 7), and in *Nasutitermes exitiosus* it is α-pinene (Fig. 8).

Fig. 6 The alarm pheromone D-limonene of the Australian termite *Drepanotermes rubriceps*

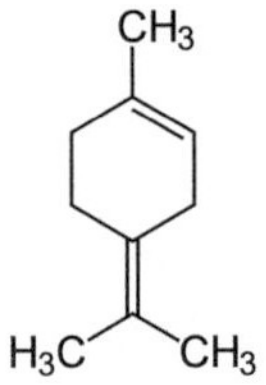

Fig. 7 The alarm pheromone α-terpinolene of *Amitermes meridionalis*

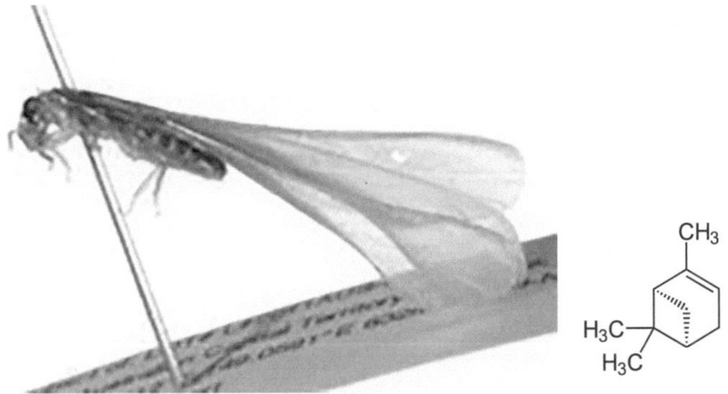

Fig. 8 The alarm pheromone α-pinene of *Nasutitermes exitiosus*

It should be noted here that terpene compounds (especially limonene and α-pinene), which serve as sex or aggregation pheromones for some bark beetles, act as alarm signals for termites. This illustrates the complex chemical interrelationships among living organisms as well as the unpredictable relationship between the chemical structure and physiological effect of the substances animals use to communicate.

Nowhere in the animal kingdom is the relationship between pheromones and hormones as clearly expressed as in social insects. Here, the pheromones of queens not only have a socializing effect but also directly influence the ontogenetic, that is, individual development of the insect. Some authors therefore refer to them as social hormones. Social insect communities are not simply colonies of cohabiting individuals, but supra-organizational units outside of which their members cannot exist. Nobel laureate Karl von Frisch said: "A farmer can have a cow, a dog, a chicken, but he cannot have a bee. Outside the hive, it perishes."

The colonies of social insects resemble a multicellular organism, whose cells can be identified with the individual insects. Just as a single cell cannot exist outside the multicellular organism, so too do socially living individuals perish outside the collective. The cells of a multicellular organism are differentiated and highly specialized to perform specific functions. Within the colony, social insects also behave in a differentiated manner and are specialized for certain tasks. In a multicellular organism, the connection between cells is mediated by the nervous system, hormones, and other chemical messengers; in insect communities, it is mediated by pheromones. In this case, pheromones are to be regarded as analogous to hormones. The insect queen can be compared to the pituitary gland of higher animals, which secretes the most important hormones to control the activity of all other endocrine glands and thus all organs.

The Chemical Signals of the Benthos

The benthos (also called benthon) represents the entirety of plant and animal organisms living above, on, or at the bottom or in the littoral zone of bodies of water. Since the beginnings of life and throughout the long evolution of the organic world, chemical signaling has remained one of the most reliable means of communication among aquatic organisms. On land, it competes with vision and hearing, but in water, under conditions of limited visibility and the density of the habitat, the advantage is entirely on its side. Aquatic inhabitants are characterized by extremely well-developed chemoreceptive organs. In lower animals, these are not yet differentiated into gustatory (taste) and olfactory (smell) senses, so chemical signals are perceived together. Higher organisms, on the other hand, possess highly specialized organs for smell and taste.

The chemical relationships between aquatic organisms are complex and diverse. Here, wireless systems operate, constantly sending and receiving information about the changing situation. Even more than on land, not only pheromones but also allomones are important in water—these are secreted by one biological species and perceived by another.

Chemical exploration helps many benthic organisms find the most suitable place for settlement. This is the case, for example, with the barnacle, which is also found on the Black Sea coast. These are small crustaceans of the genus *Balanus* (Fig. 1).

They live hidden in small white conical shells and are fused at their base with various solid materials such as mussel shells, crabs, coastal rocks, etc. In summer, the barnacle lays a large number of eggs, from which

I. G. Ivanov, *The Invisible Language of Nature*,
https://doi.org/10.1007/978-3-662-73302-8_9

Fig. 1 *Balanus* (barnacle). (© medveh/Getty Images/iStock)

free-swimming juveniles hatch. At a certain stage of development, the larvae attach themselves to a solid substrate to transform into adult crustaceans after metamorphosis. However, before settling, the larvae must carefully select the site. They explore the surface with their antennae, which have two discs at the end covered with fine cilia. A place where another barnacle has lived is considered suitable. The logic is simple: "If this place nourished my ancestors, it will nourish me as well." The traces left by the ancestor are proteins similar to those of the chitinous shell of crustaceans. Their peculiarity is that they are absolutely insoluble in water. When the larva examines the site, it therefore analyzes not substances dissolved in water, but solid substances. In doing so, it reliably distinguishes its own proteins from foreign ones. Exactly how the barnacle achieves this is hard to say. However, it can be assumed that the sensitive cilia of the antennae are capable of recognizing the configuration of protein molecules fixed on the surface. In this case, the antennae may not serve as olfactory organs, but rather as hypersensitive fingers with which the larva can "feel" individual molecules.

Some freshwater crabs are cannibals. The adults eat the young. For this reason, the larvae attach themselves to their mother's body for safety. If a larva falls off, it searches for the mother and finds her unerringly among many other female crabs. If it does not find her, it seeks protection with another female carrying larvae on her body. The only way to find a foster mother is through the maternal pheromones she secretes.

Pheromones are also important for the lives of organisms such as corals, sea anemones, etc. Sea anemones, also called actinias, usually reproduce sexually and less often vegetatively (by budding). The sperm enter the female's interior with the water and fertilize her eggs there. The tiny, ciliated eggs grow to a certain stage in the mother's internal cavity and are then released. They then attach themselves to a hard object and begin their sessile life. Budding of sessile actinias produces a colony genetically identical to the original.

Sea anemones have the remarkable ability to distinguish their own species from others. *Anthopleura elegantissima* (Fig. 2) inhabits the rocks of the North American west coast and tolerates no foreign anemones. If one approaches, war is immediately declared as follows: The stinging threads located next to the crown of tentacles swell and form bladders called acontia. These take the shape of fingers and are directed at the foreign anemone. The acontia contain toxin-filled cells that, upon contact, discharge into the enemy's body. This forces the foreign anemone to detach from the rock, or else it will be killed. The fight lasts only a few seconds and ends with the victory of one of the combatants. How can the anemone, which can neither see nor hear nor has a brain, distinguish its own species from others? Undoubtedly, recognition occurs through chemoreception, but whether the mediators are low- or high-molecular substances is unknown. The American researcher L. Francis observed that the formation of acontia is preceded by the touching of the tentacles of the two "hostile" anemones. It is likely that the anemones recognize foreign proteins in a manner similar to the way barnacle larvae do. The "xenophobia" of anemones leads to the formation of an empty strip of no-man's-land between neighboring heterogeneous colonies. L. Francis conducted an interesting experiment. He divided an aquarium with a glass partition and populated the two chambers with colonies of unrelated anemones. After the population reached a certain density, he removed the partition so that the different colonies were in close proximity. Immediately, war

Fig. 2 Sea anemone *Anthopleura elegantissima*. (© noblige/Getty Images/iStock)

broke out, quickly leading to the formation of a neutral zone. Interestingly, anemones only wage war with their own kind, not with their prey or enemies. While anemones recognize each other by direct contact, enemies detect them from a distance. The anemone described above, for example, often falls victim to the broad-barred aeolid nudibranch *Aeolidia papillosa* (Fig. 3).

This is a 4–6 cm long predator whose body is covered with numerous pink, flat papillae. It feeds exclusively on anemones, with its favorite prey being the *Anthopleura* genus. However, it does not eat the entire individual, but only its tentacles. Injured anemones are particularly attractive. The presence of an injured anemone triggers a strong reaction in the other anemones of the colony. They contract convulsively, roll up their tentacles from all directions, then pull them inward, transforming into a dense, slimy mass that is several times larger than the expanded anemone. Clearly, injury to an anemone is associated with the release of an alarm substance that signals imminent danger. American researchers N. Nove and J. Sheikh set out to isolate and identify this alarm pheromone. They named it anthopleurin. It turned out to be a quaternary amine with a betaine structure (Fig. 4). Even at very low concentrations (3.5×10^{-10} mol/l), it causes convulsive reactions in anemones. At the same time, it acts as an attractant for *Aeolidia* (broad-barred aeolid nudibranch).

Anthopleurin is the second pheromone in invertebrates (after crustecdysin in crustaceans) with a precisely defined chemical structure.

Where does the anthopleurin in *Aeolidia* come from? It could originate from the consumed anemone or be synthesized in *Aeolidia* itself. Studies have shown that anthopleurin ingested with food is not altered by *Aeolidia*.

Fig. 3 *Aeolidia papillosa*. (© Sakis Lazarides/Getty Images/iStock)

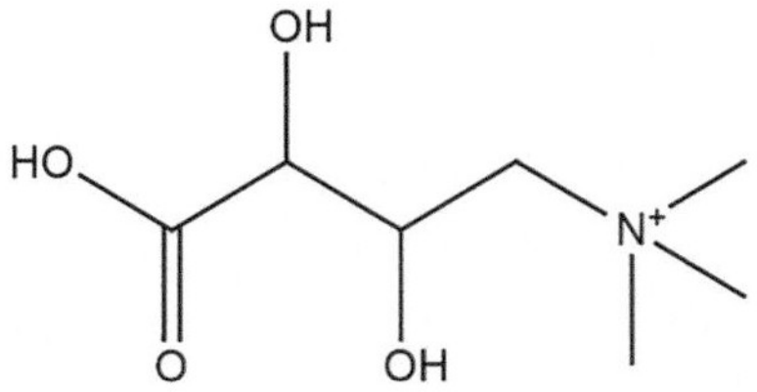

OH
HO
N+
O OH

3-carboxy-2,*3-dihydroxy-N,N,N* ,-trimethylpropane-1-aminium

Fig. 4 Anthopleurin

It accumulates in its tentacles. As a result, the predator fed with anemones begins to "smell" and frighten other anemones. Since the concentration of anthopleurin in the tentacles of *Aeolidia* exceeds the threshold for effect on anemones by a factor of one hundred, it is still released days after the feast. Analysis of various anemone tissues shows that the anemone stalk contains 4 to 5 times more anthopleurin than the tentacles, which in turn explains why *Aeolidia* prefers to eat only the anemone tentacles. If the stalk is also eaten, *Aeolidia* absorbs so much anthopleurin that it must fast for weeks until it "dies."

The exchange of chemical information is also the basis of the symbiosis between anemones and other marine organisms. It is known that anemones feed on fish, crabs, shrimp, snails, etc., first paralyzing them with their toxic tentacles and then swallowing them. However, some fish species get along excellently with anemones and act for mutual benefit. They live in symbiosis. This includes, for example, the clownfish (Fig. 5), which hides from its predators among the anemone's tentacles and at the same time protects its host from enemies.

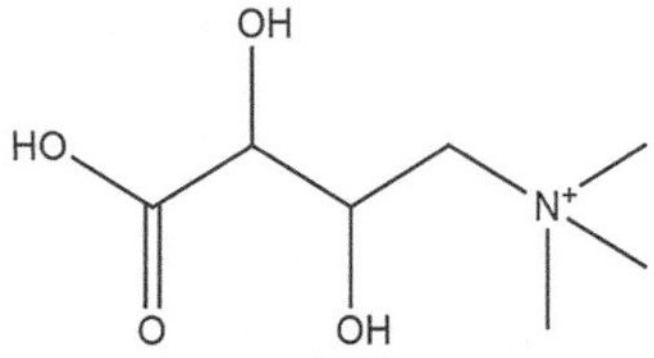

OH
HO
N+
O OH

3-carboxy-2,*3-dihydroxy-N,N,*N-trimethylpropane-1-aminium

Fig. 5 Clownfish, lives in the tentacles of *Anthopleura elegantissima*. (© JodiJacobson/Getty Images/iStock)

Excretions from the clownfish can benefit the sea anemone. Fish of this species are loyally devoted to their protector and do not leave her. It has been shown that the "landlady" is identified by her "scent."

Pheromones regulate the population size of many aquatic organisms. The spotted planarian *Dugesia tigrina* (Fig. 6), for example, reproduces very rapidly. However, when the population density reaches a certain threshold, it secretes a pheromone that inhibits the activity of the nerve cells at the worm's head end, which control reproduction. As a result, the worm stops dividing.

In the planarian *Phagocata gracilis*, sexual pheromones act as aggregation pheromones. Their secretion causes both males and females to come together. Egg-laying begins only when at least 200 individuals have gathered in a group.

Chemically, the sexual pheromones of aquatic organisms, in contrast to those of insects, are only poorly studied. Little is also known about their physiological effects and their role in the mating behavior of animals. In most cases, they act both as sexual and aggregation pheromones, thus helping to bring together a large number of heterosexual individuals in one place. This enables the simultaneous release of seminal fluid over a small area, which in turn greatly increases the efficiency of fertilization. However, it should be noted that the simultaneous release of sperm and eggs in one place does not necessarily mean fertilization will occur. In the aqueous medium, the male seminal fluid is quickly diluted, reducing the likelihood of random encounters between sperm and eggs. This is where gamones

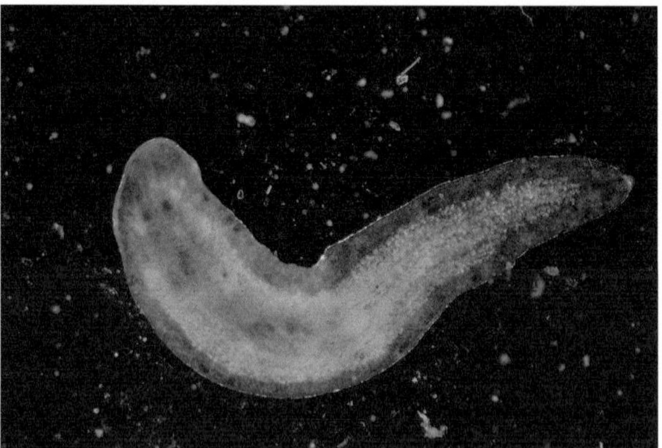

Fig. 6 Spotted planarian *Dugesia tigrina*. (© Sinhyu/Getty Images/iStock)

(pheromones of gametes) come into play, which are secreted by the eggs and serve as attractants for the sperm.

Gamones not only attract sperm, but also their accumulation, so-called spermatophores, such as those of the Argonautidae (paper nautiluses [octopuses]). The spermatophores travel great distances in the water to find a receptive female and fertilize her. Since this case is unique, we will examine it in more detail. Female and male Argonautidae differ like giant and dwarf. While the female can reach up to 30 cm in length, the males grow only a few centimeters. During the breeding season, a special organ of the male paper nautilus ("spermatophore pouch") produces seminal fluid, which is packed into "packets" called spermatophores. The structure of the spermatophore is complex and resembles a sea mine. It contains a "detonator wire," a tightly wound protein spring connected to a biological "fuse." When the mating season begins, one of the octopus's tentacles grows, grabs one of the spermatophores, and detaches it from the animal. By deforming its body, the small torpedo sets off in search of a receptive female in the depths of the sea. The only signal it can follow is the specific chemical signals emitted by the female. The torpedo enters the female's mantle cavity, the biological fuse is triggered, and the spring ejects the packaged sperm with full force. The detached tentacle with the spermatophore is so mobile that it was long considered an independent animal. The French naturalist Georges Cuvier (1769–1832) named such a tentacle hectocotylus (Fig. 7).

Chemical signals are also important for fish reproduction. Here, too, sexual pheromones contribute to mating. If such a pair is encountered by a fish of the same species, it triggers aggression in the male mating partner, even if he is in an opaque, semi-permeable cellophane bag. The *Bathygobius* male (goby) begins to guard its territory shortly before spawning, and when a

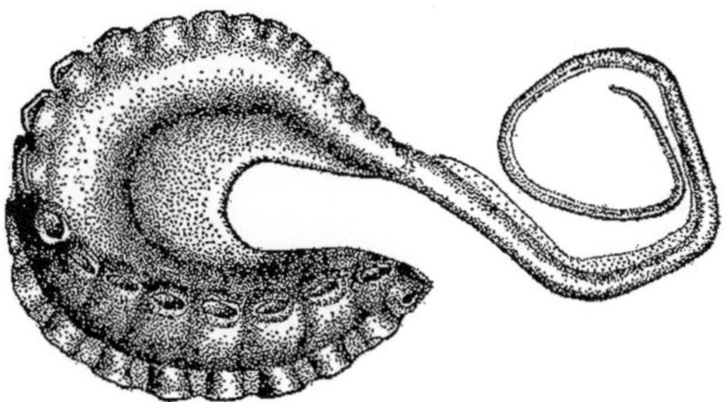

Fig. 7 Hectocotylus

sexually mature female appears, it starts attacking potential rivals. The same behavior is observed in an aquarium after water from another aquarium containing a sexually mature female is added. It has been shown that the sexual pheromone in this case is secreted by the urogenital system and is present in the urine, blood, and eggs of the female.

"Trouble!" in the Language of Fish

While little is known about sexual pheromones in fish, their alarm pheromones are among the best-studied in vertebrates. Their discovery is associated with the Nobel laureate Karl von Frisch. In 1938, he trained a school of wild minnows (*Phoxinus phoxinus*) to approach the riverbank when a bell rang. He then fed them worms. On one occasion, he caught a fish and performed a minor surgical procedure—he injured the sympathetic nerve, which controls the chromatophore muscles, with a needle. Chromatophores are skin cells near the tail that can change their color. After releasing the operated fish, von Frisch noticed that the other fish in the school swam away and did not return for hours. He assumed that they were frightened by the altered appearance or that he had somehow managed to inform their companions of the improper act of this seemingly benevolent man. To rule out the possibility of exchanging information with the other fish, von Frisch killed the animals after the operation and then threw them into the water. The reaction was the same. Then he cut them into pieces so that they no longer looked like fish. When he threw the pieces into the water, the school scattered again. It was obvious that the panic reaction had nothing to do with the altered appearance of the fish. He suspected that the injury had led to the release of a substance that frightened the other fish. To prove his hypothesis, von Frisch prepared an aqueous extract from crushed fish, which he filtered to remove all traces of animal tissue. When the extract was added to the water, the fish again swam away. There was no doubt that the surgical procedure was accompanied by the release of an alarming substance that

I. G. Ivanov, *The Invisible Language of Nature*,
https://doi.org/10.1007/978-3-662-73302-8_10

triggered a fear response. According to modern terminology, this is an alarm pheromone.

The next question von Frisch had to answer was where the production of the alarm substance takes place. To answer this, he examined the ability of various organs (intestine, liver, skin, etc.) to trigger a fear response. Only the skin had a positive effect. Thus, the alarm substance is contained in the skin and is released when it is injured. This raises further questions: Which part of the skin exactly releases the alarm pheromone? Do fish, like insects, have specialized organs for the production of alarm pheromones, or is the pheromone produced by some ordinary component of the skin epithelium?

Karl von Frisch's students were involved in researching fear pheromones and the fear response of fish. Through careful histological examination of the skin, they found that, in addition to the long-known secretory cells that produce skin mucus, the skin also contains another type of secretory cell in the form of bulbs and rods. While the former open up and release their secretion onto the skin surface, the latter are located deeper and have no direct contact with the surface of the skin epithelium. These are the cells that produce the mysterious alarm substance. The discovery of these cells also explains why alarm pheromones are only released when the skin is injured. Unlike insects, fish cannot sound the alarm on their own. False alarms are also impossible for them. Here, every danger signal is real and means: "Run and save yourself!" With the help of the alarm pheromone, even a fatally wounded fish can warn its conspecifics of impending danger. And if a predator has had the opportunity to injure and eat a fish, it will have to wait quite a while for the next one.

After the fear response and the presence of alarm pheromones had been demonstrated in the minnow, they were also sought in fish of other genera and families. The fear response was observed in most freshwater fish studied and in comparatively few saltwater species. Researchers also looked for a correlation between the presence of fear responses and the lifestyle of the fish (whether they live in groups or not). However, nothing was found.

However, all fish that show a fear response have rod cells in their skin. This proves their relationship to alarm pheromones beyond doubt. In different fish species, they vary in appearance and size, but are always located beneath the skin surface.

Karl von Frisch and his colleagues also investigated the species specificity of the alarm substance, i.e., whether it is recognized only by fish of the same species or whether other species also respond to it. The researchers found: The alarm pheromone is species-specific! In rare cases, it is also recognized by individuals of closely related species, but their response is significantly weaker than that of fish of the same species. Although the fear response is innate, it is absent in newborn fish. Their skin already produces alarm pheromones. Their sense of fear develops between the 20th and 60th day of life.

What is the chemical nature of the alarm substance? Although the effect of the alarm pheromone is well studied, its chemical formula has still not been conclusively determined. It has been found that the alarm substance fluoresces when exposed to ultraviolet light. Based on this, methods for fluorescence microscopic observation of the producing cells have been developed. According to most researchers, the alarm pheromone is a pterin. Pterins are a large class of organic compounds, so named because they were originally found in the wings of butterflies (pterygium, the wing). Their structure is similar to that of purines, which are part of nucleic acids. They are also biogenetically related to purines. In 1977, the alarm pheromone of minnows was identified as 6-dihydroxypropyl-iso-xanthopterin and named ichthyopterin (Fig. 1). However, this structure is disputed.

Another interesting pterin is neopterin (Fig. 2). Its structure is similar to that of ichthyopterin. It is synthesized by human macrophages (which belong to the white blood cells) under the stimulating effect of gamma-interferon and is used in clinical medicine as a marker for the activation of cell-mediated immunity.

Fig. 1 Minnows (*Phoxinus phoxinus*) and their alarm pheromone ichthyopterin. (Photo: © CreativeNature_nl/Getty Images/iStock)

Fig. 2 Neopterin

Hypoxanthine-3-N-oxide is another possible alarm pheromone. As a chemical compound, it has a deterrent effect ("Schreckstoff" according to von Frisch), but it is not found in fish skin. A more recent study has shown that chondroitin (a type of glycosaminoglycan) could also be an alarm pheromone. It triggers an alarm response at low concentrations and is capable of stimulating the olfactory bulb in the brains of experimental fish. Most likely, the alarm pheromone in fish is a complex mixture of several biologically active substances that reinforce each other.

Regardless of their chemical nature, fish pheromones are strong repellents (serving to ward off pests or intruders) with a very low threshold concentration. To study their effect, Karl von Frisch introduced a standard procedure for preparing fish skin extracts in the 1940s. It is still used today and is very simple: 200 mg of fresh skin are cut into pieces with scissors and extracted in 200 ml of water. After 30 minutes, the extract is filtered and used immediately. The repellent effect of extracts prepared in this way is very strong and can still be detected at a 50,000-fold dilution. 0.002 mg of skin in a 14-liter aquarium is enough to trigger a fear response. The threshold concentration of the crudely purified pheromone is 2.5×10^{-6} µg/ml. These data explain why a single needle prick is enough to scare off minnows. They also explain why a fish that is released from a fishing hook and falls back into the water will prevent other fish from biting again for hours.

How Do Fish Perceive Alarm Signals?

The answer to this question was again provided by Karl von Frisch. He demonstrated that fish with experimentally impaired sense of smell do not react to the alarm substance and show no fear response. Thus, alarm pheromones are perceived via the sense of smell. For us, as air-breathing creatures, it is difficult to imagine that fish can also have a sense of smell. The sense of smell is essentially a form of remote perception. Its organs respond to chemical signals, regardless of whether they are dissolved in water or air. Like us, fish have noses, but their mucous membrane is surrounded by water, not air.

The olfactory organ of fish is a U-shaped tube located directly in front of the eyes and has two openings—a nostril and an outlet (Fig. 1).

Fish are among the animals with the most highly developed sense of smell. Their olfactory acuity may even rival that of dogs. Based on this sense, complex relationships are established both among individuals of the same species and with the environment. Numerous experiments have shown that fish are able to recognize the sex, age, and other individual characteristics of the members of the school in which they live. According to the American marine researcher K. Todd, chemical signaling in fish is so refined that they can even transmit information about their mood. He conducted long-term experiments with ratfish (a cartilaginous fish) and observed that even in overcrowded aquariums, these fish could live together peacefully. However, if they quarreled and water from an aquarium where the animals were calm was added to the tank, the fighting quickly stopped. Conversely, agitation was transferred if water from a restless aquarium was introduced into a peaceful one. K. Todd describes the following curious case: In one aquarium,

I. G. Ivanov, *The Invisible Language of Nature*,
https://doi.org/10.1007/978-3-662-73302-8_11

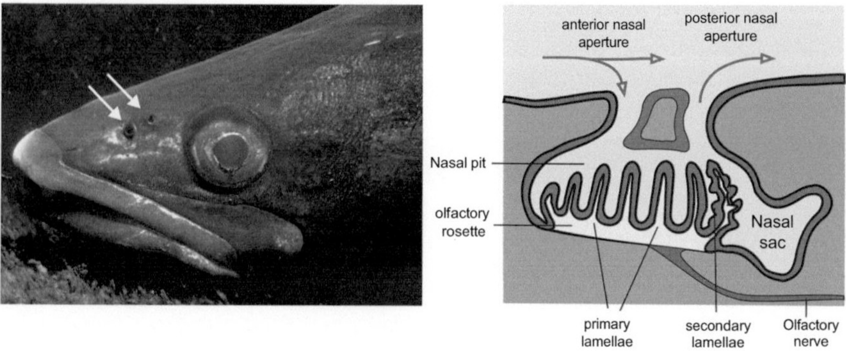

Fig. 1 Olfactory organ of a fish

several juveniles and an adult ratfish lived together. However, the adult was aggressive and constantly chased the juveniles, so they were always restless in its presence. When Todd moved the adult to another aquarium, the juveniles quickly calmed down. But when he poured water from the adult's tank into the tank with the juveniles, panic broke out immediately.

Fish also transmit information about their social rank through chemical signals. It has been experimentally demonstrated that a dominant fish in the hierarchy retains its supremacy if it is separated from the aquarium for a period and then returns. However, if it is placed in an aquarium with other, unfamiliar fish and is defeated in a fight, it loses its rank upon returning to the original aquarium. Then, even fish that were previously clearly subordinate in the hierarchy attack it. This behavior, however, is not observed if the fish are deprived of their sense of smell.

Fish also use their sense of smell to find their birthplace. This is one of the most astonishing phenomena in the animal kingdom—the migration of fish.

Many migratory fish are known, but salmon and eel are the leaders in terms of the extent of their migrations. Both species have two lifestyles. Salmon spawn in freshwater and live in the oceans, while eels, in contrast, spend their lives in freshwater and spawn in the salty waters of the Sargasso Sea. The migration of the Pacific salmon (Fig. 2) has been studied in the greatest detail.

Salmon spawn from November to December in fast-flowing, clear rivers in the north, and the young fish hatch in spring. At first, they are colorless and translucent with black eyes. Pigmentation occurs toward the end of summer. After about two years, they reach a length of 15 cm and take on the coloration of adult fish—blue back with light sides and dark longitudinal stripes. Only in the third or fourth year does the salmon go to sea, where it

Fig. 2 Pacific salmon. (© Supercaliphotolistic/Getty Images/iStock)

spends most of its life. After reaching sexual maturity, it returns to its native streams, where it spawns and then dies. Using the most advanced methods, it has been shown that the salmon spawns precisely at the spot where it hatched. To get there, it overcomes incredible obstacles—it moves upstream in raging rivers, climbs waterfalls, and evades predators. And all this in the name of duty to its ancestors—to pass on its genes.

The journey of the salmon borders on miraculous when it comes to its orientation and navigation. Not only is it an excellent navigator, but it also has an outstanding memory, which recalls the features of its birthplace over many years. How do salmon find their birthplace? It has been discovered that they use two navigation systems for this purpose—the visual (solar) and the sense of smell. The former brings them to the mouth of their home river, the latter to the very spot where they hatched.

The hypothesis that salmon use their sense of smell for orientation was proposed by A. D. Hasler. He supported his assumption experimentally with two types of trials. The first aimed to demonstrate the acuity of the salmon's sense of smell. The second, their ability to orient themselves toward their native habitat using smell. In the first type of experiment, Hasler found that salmon quickly develop a conditioned reflex to odors. When a salmon smells a particular substance, it can easily distinguish it from many other substances with similar odors. Thanks to its highly developed sense of smell, it can detect aromatic substances at a concentration of 3×10^{-18} g/l. This is roughly equivalent to a bottle of perfume diluted in the Black Sea.

After demonstrating the sensitivity of the salmon's sense of smell, Hasler also showed their ability to locate their birthplace by scent. To do this, he released the salmon several kilometers downstream from a river fork. He found that the fish swam upstream without hesitation, directly into their stream. However, when he blocked their olfactory openings, they were no longer able to find their birthplace. From this, he concluded that the visual navigation of salmon is only necessary to find the correct river, and that olfactory orientation takes over thereafter. When placed in their home river, the salmon still found the exact spot even when they were blinded. For Hasler, this was conclusive evidence that salmon do not need their vision to find their native site.

Interesting results supporting Hasler's conclusions were also obtained from electrophysiological studies of the salmon's olfactory organs. When the olfactory organ was rinsed with water from the hatching site, the olfactory bulb, located at the base of the brain, showed strong electrical activity. Water from other sources did not trigger a similar response. Thus, the water from the birthplace carries a specific scent that may have imprinted the salmon during the early stages of embryonic development. It is a kind of composite odor, resulting from the minute amounts of organic substances released by the plants and animals living there. Their concentration in the water is so low that it is impossible to detect them by chemical methods and yet it far exceeds the threshold concentration required to stimulate the salmon's olfactory receptors and trigger a physiological response.

Olfactory navigation has not been conclusively demonstrated in eels. However, since their sense of smell is in no way inferior to that of salmon, it is assumed that it serves as an orientation aid in coastal marine waters.

The sense of smell in sharks is also astonishing. The scent of fresh blood is a powerful attractant, which they can detect from miles away. To avoid attracting sharks, the golden rule for divers is not to have injured fish nearby and not to swim with a bleeding wound. Since sharks pose a deadly threat to shipwreck survivors and beachgoers in many countries, effective shark repellents have long been sought. Special shark repellents are available, but unfortunately, they are not reliable enough. One such repellent is the American "Anti-Shark," which was issued to American sailors and military pilots during World War II. It is a tablet made of copper acetate and a dye that colors the water dark blue. After the war, the effect of the Anti-Shark agent was studied more closely, but its protective effect could not be confirmed. The famous oceanographer and shark researcher Jacques Cousteau (1910–1997) joked: "Sharks like to eat it, but apparently it does not improve their digestion," i.e., its effect during the war was purely psychological.

The Living Gas Analyzer

Immersed in our daily lives, we hardly notice the smells around us. We do not notice the blooming roses, the scent of the pine forest, the smell of freshly cut hay, the aroma of fruits and vegetables on our table. Yet it is enough to direct our attention to them, and we immediately realize that we literally live in a sea of odors. We are confronted with hundreds, even thousands of them every day. There is the smell of tea, coffee, soup, gasoline, creams, perfumes, and so on. We perceive and analyze them unconsciously. Our awareness of odors becomes conscious when they are very strong, pleasant, or unpleasant, and when we focus on them.

For modern humans, the sense of smell is of limited importance compared to the other senses and serves primarily to assess the comfort of our living environment and the taste of food. In the latter case, it is complemented by perception. For our ancestors, however, the sense of smell was vital. By scent, they could detect a wild animal hidden in the jungle or the approach of a predator. And although the sense of smell has atrophied over the course of evolution, it remains one of the most sensitive human senses even today. Every developer of gas analyzers envies the ability of our nose to detect odors. Humans are capable of perceiving the smell of one trillionth of a gram of skatole (a substance found in excrement), five ten-millionths of a gram of vanillin, four hundred-millionths of a gram of ethyl mercaptan, and so on. The average person can distinguish between 2000 and 3000 types of odors. The organic chemist and the specialized perfumer can recognize significantly more—up to several tens of thousands of odors. When the other senses are impaired, the sense of smell compensatorily increases

I. G. Ivanov, *The Invisible Language of Nature*,
https://doi.org/10.1007/978-3-662-73302-8_12

its sensitivity. Literature describes cases in which blind and deaf people have recognized family members, relatives, and acquaintances by their scent.

The sense of smell in many mammals is even far more sensitive than that of humans. A dog, for example, can detect the scent of butyric acid at a concentration of 9000 molecules per cubic centimeter. For reference: 1 cm^3 of air contains 26,800,000,000,000,000,000 gas molecules. However, it is not clear whether the dog can distinguish as many odors as humans, or whether this hypersensitivity applies only to certain substances.

How Do Humans and Mammals Perceive Odors?

This question has two aspects. 1. How do chemicals interact with olfactory receptors to convert chemical signals into nerve impulses? 2. What is the relationship between the chemical structure of substances and their odor?

The olfactory system of mammals and humans consists of two sections—the peripheral and the central part (Fig. 1). The peripheral part is located in two groove-like indentations in the nasal cavity, while the central part is in the brain. The indentations are covered by the olfactory mucosa. This is a slimy, yellow-brown tissue with a total area of 3–5 cm^2. In different mammals, the area varies according to their body size. The olfactory epithelium consists of two types of cells—supporting cells and olfactory cells.

In fish, for example, 1 mm^2 of the olfactory epithelium contains 40,000–60,000 olfactory cells, while in rabbits and dogs it is 200,000. Olfactory cells are specialized nerve cells. They are spindle-shaped with a cross-section of 5–10 μm (1 μm = 1 millionth of a meter = 10^{-6} m).

Each cell has two ends, one of which terminates in cilia, 0.1 or 1–2 μm thick. The second, the axon, leads directly into the brain. The cilia of the olfactory cells are located above the epithelial surface and float in mucus secreted by a special gland, the Bowman's gland. Due to the small cross-section of the cilia, their surface area is enormous. In humans, it amounts to 600 cm^2. This is where the contact between the molecules of odorants and the nervous tissue takes place. The axons of the olfactory cells, which are bundled in groups of 20–100 within a common Schwann sheath, connect

I. G. Ivanov, *The Invisible Language of Nature*,
https://doi.org/10.1007/978-3-662-73302-8_13

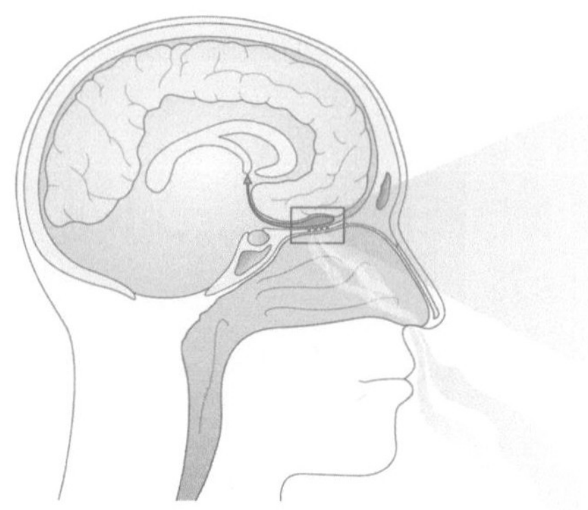

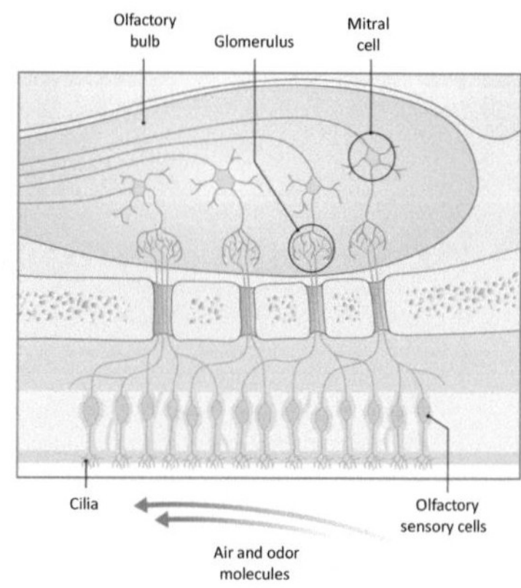

Fig. 1 Human olfactory system

to the olfactory bulb, which is located in the anterior part of the brain. Its shape can be round or oval, and its size varies from species to species. The size of the olfactory bulb indicates the importance of smell for the respective animal species. In some animals (e.g., kangaroos), it occupies half of the two hemispheres of the forebrain. In dogs, the relative proportion of the olfactory bulb is much larger than in humans. In mammals, the olfactory bulb consists of two symmetrically arranged parts, with the axons from the left nostril entering the left part of the brain and those from the right nostril entering the right part.

Considering the structure of the olfactory system, the short path from the receptor, which interacts directly with chemical substances, to the brain is striking. The brain analyzes the incoming information. In vision and hearing, the path is longer. In the eye, there are a cornea, a lens, and a vitreous body between the retina and the external environment; in the ear, a tympanic membrane and auditory ossicles. The Canadian scientist R.W. Wright, a specialist in chemoreception, writes: "There is only one synapse between the primary sensory tip of the olfactory epithelium and the olfactory centers of the brain. A closer connection between the organism and the environment is hardly imaginable." And Wilfrid Le Gros Clark says: "… from an evolutionary perspective, the olfactory bulb, in which the olfactory fibers terminate, represents a part of the brain that has been relegated to the periphery, and the direct connection of individual receptors with it is an expression of the fact that the vertebrate brain originally developed primarily as an organ of smell."

Obviously, the sense of smell is much more important for most animals than vision and hearing. Even in humans, for whom it is of secondary importance, the number of axons originating from the olfactory cells is about 100 million, while the optic nerve has 1 million and the auditory nerve 800,000.

Electrophysiological studies have made significant contributions to the investigation of the functioning of the olfactory apparatus. The method makes it possible to track changes in electrical potentials even in a single olfactory cell. It has thus been shown that olfactory receptors in mammals are not uniform. Under the influence of a particular chemical, some of them are more strongly excited, others less so, and still others do not react at all. The latter, in turn, are excited by other chemical stimuli. In other words, there are more or less specialized receptors in the olfactory epithelium. The number of different olfactory cells is not known, but can be roughly estimated. Let us assume that two types of neurons are involved in the perception of odors and that each of them has two states, namely an excited (+) and a non-excited (–) state. Thus, there can be 4 (2^2) different signals. We usually denote them as (– +), (– –), (+ –), and (+ +). There can be 8 signals with two neurons and 2^n signals with n neurons. If a person is able to distinguish 10,000 odors, then $2^n = 10{,}000$, from which it follows that $n = 13$. Since nature certainly provides a certain information reserve, the number is probably higher.

The information originating from the olfactory cells is sorted in the olfactory bulb, and the signals from individual receptors are summed. It is assumed that the mitral cells, which connect the olfactory bulb with the olfactory centers of the brain, transmit the summed impulses of receptors of the same type. The mixing of signals, i.e., the integration of information and the formation of the sense of smell, then takes place in specialized brain regions.

How Does the Odor Molecule Interact with the Odor Receptor?

From here on, we enter the world of hypotheses and conjectures. The basis for excitation of the olfactory receptor, as with any other nerve cell, is the depolarization of the cell membrane. Let us recap: Every cell contains potassium ions, while the extracellular space contains sodium ions. Thus, there is a significant potential difference between the inside and outside of the cell membrane. If external influences alter the membrane's permeability to these two types of ions, their concentrations on both sides of the membrane also change. This leads to an immediate change in the electrical potential. The electrical signal thus generated propagates as a nerve impulse along the length of the neuron. Two states are possible for the neuron: polarized (signal "+") and depolarized (signal "−"). Intermediate states are not possible.

How do chemicals depolarize the membranes of olfactory cells? For a substance to be perceived as an odor, it must meet several conditions: a) it must be volatile, i.e., it must be able to maintain a certain concentration in the air through which it reaches the nose; b) it must have a balanced solubility in water and organic solvents; c) it must have a moderately large molecular mass. Substances with a molecular mass of more than 300 daltons are considered odorless. The second condition mentioned above is related to the fact that the cilia of the olfactory receptors are immersed in an aqueous medium, and contact with them is only possible if the odorant diffuses into the mucus. Higher paraffinic hydrocarbons, which are insoluble in water, are

I. G. Ivanov, *The Invisible Language of Nature*,
https://doi.org/10.1007/978-3-662-73302-8_14

odorless. However, odorants must also exhibit a certain degree of lipophilicity, since olfactory receptors contain hydrophobic functional groups required for van der Waals interactions.

Experiments with labeled compounds have shown that the molecules of odorants undergo chemical changes after adsorption onto the olfactory epithelium, i.e., they are metabolized. It is remarkable that even compounds as chemically inert as hydrocarbons are converted at the receptor into carbonyl compounds and organic acids. However, these transformations are only possible in the presence of oxygen. When subjects were presented with aromatic substances in an oxygen-free gas mixture, they perceived no odor. Oxygen is therefore necessary for our sense of smell. It is unclear whether the observed biochemical reactions are an intermediate step in the generation of the nerve impulse or whether they are required to remove the odorant from the receptor, to cleanse it and prepare it for a new act of reception. From all that has been said so far, it is clear that the processes in the olfactory epithelium are extremely complex and have only recently been elucidated. More than 30 hypotheses, some of them quite exotic, have been proposed to explain these processes.

The science of odors, known as odorology, only gained a solid scientific foundation at the beginning of the twenty-first century, when Richard Axel and Linda B. Buck were awarded the 2004 Nobel Prize in Medicine for their discovery of G-receptors in olfactory cells. According to them, the sense of smell is a chemical stimulus with a physiological response triggered by the interaction of one or more molecules via G-proteins associated with olfactory receptors. G-proteins are transmembrane proteins with 7 alpha-helices, embedded in the membrane of the olfactory cell such that one end (N-terminus) is outside the cell and the other (C-terminus) is inside the cell (Fig. 1). This enables interaction with molecules from the extracellular space and transmits this interaction to intracellular structures. To investigate the structure of G-proteins, Axel and Buck used genetic engineering methods. They succeeded in isolating the genes of 18 different G-proteins and expressing them in bacteria to obtain sufficient protein for structural studies. According to the theory of Axel and Buck, after the odorant molecule binds to the extracellular domain of the G-protein, structural changes occur that are reflected in the intracellular domain.

Odorants from the air pass through the mucus directly or via transport proteins and reach the olfactory receptor, which undergoes structural changes and activates G-proteins. An active subunit (Ga) is released inside

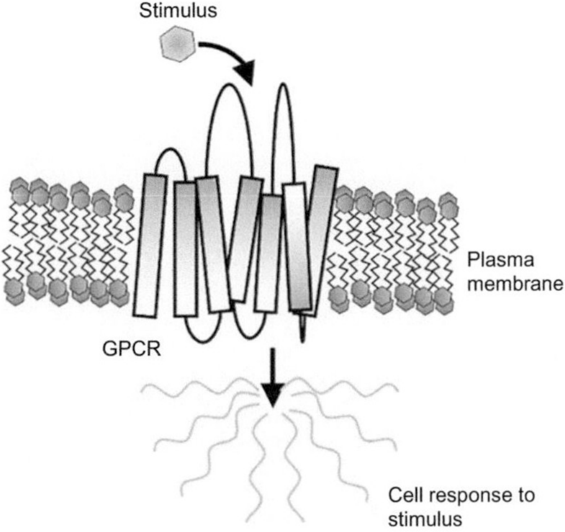

Fig. 1 G-protein olfactory receptor (GPCR)

the cell. This activates adenylate cyclase, leading to the conversion of adenosine triphosphate (ATP) into cyclic adenosine monophosphate (cAMP). The cAMP is able to open cyclic nucleotide-gated ion channels, allowing Ca^{2+} and Na^{+} ions to enter the cell. This leads to depolarization of the olfactory receptor neuron and generates an electrical impulse, which is transmitted from the axon of the olfactory neuron to the brain (Fig. 2).

Figure 3 illustrates Axel and Buck's concept of how three different volatile substances interact with three different G-receptors. This results in the perception of different odors. The fact that cAMP also mediates the response of certain hormone groups suggests a phylogenetic relationship between hormones and pheromones. In the early stages of the evolution of life, when there were no multicellular organisms but only colonies of unicellular organisms, communication between cells occurred via chemical signals. These first signals can be regarded as precursors of hormones (pheromones).

It is likely that some of the mechanisms of early chemoreception have been preserved to this day and have evolved into more precise hormone and pheromone reception.

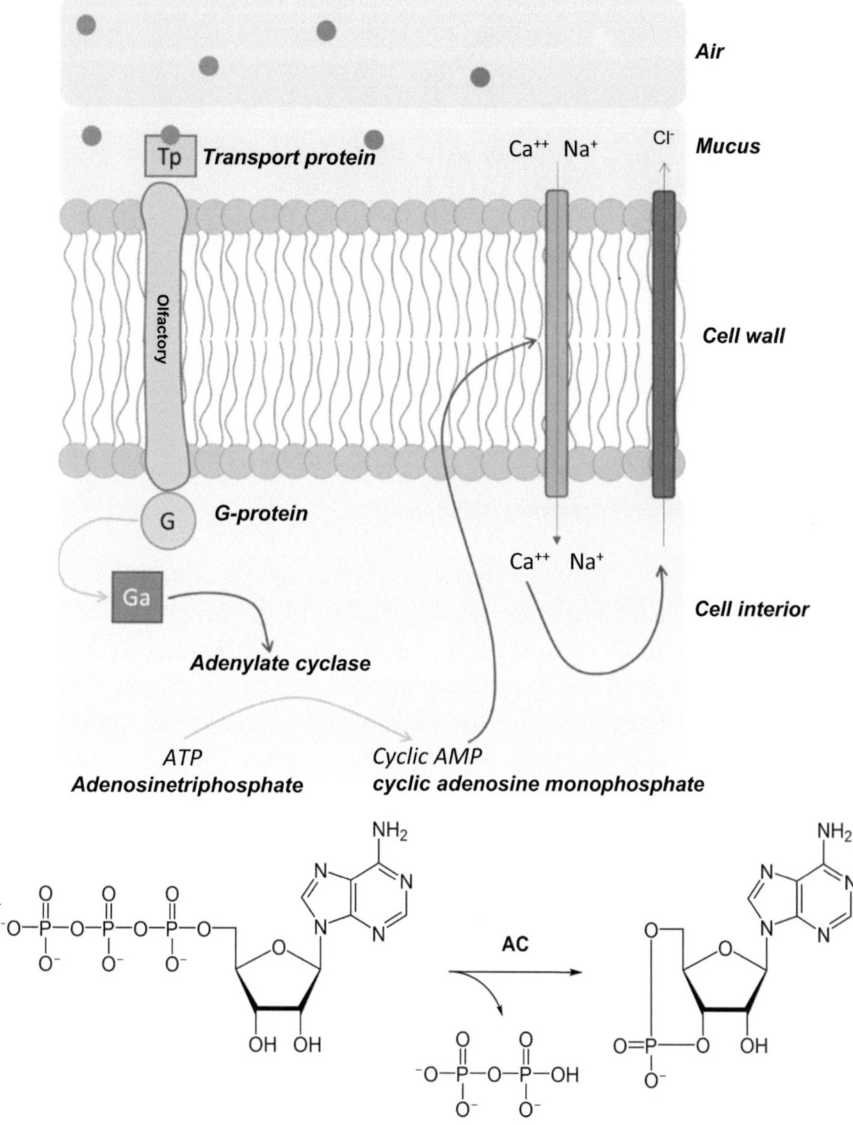

Fig. 2 Reaction cascade after the entry of an odorant into the nasal cavity

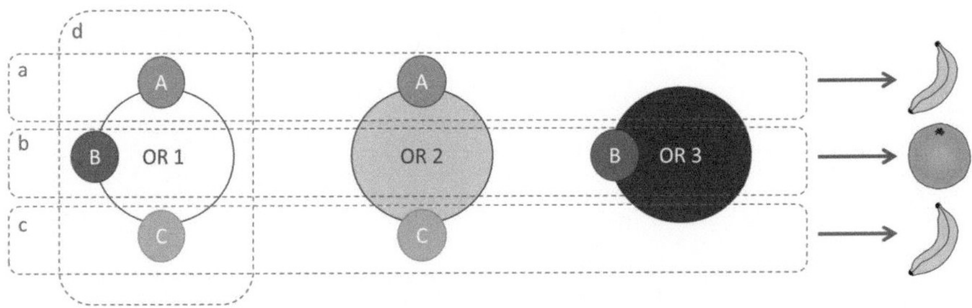

Fig. 3 Response of olfactory receptors to odors. OR1, OR2, and OR3 are three different olfactory receptors; A, B, and C are three different odorant compounds. Each of the receptors can bind one or more odorants. In the case shown, the OR1 receptor can bind all three molecules (A, B, and C), the OR2 receptor binds only A and C, and the OR3 receptor only B. The following odors result from these combinations: **a**) OR1-A + OR2-A – banana scent; **b**) OR1-B + OR3-B – scent of orange; **c**) OR1-C + OR2-C – banana scent

Molecules and Odors

The question "What determines the smell of a chemical compound?" has fascinated scientists for a long time. To this day, however, there is no definitive answer. The rapid development of organic chemistry at the end of the nineteenth century and the desire of chemists to synthesize new aromatic compounds for the perfume industry led to the first hypothesis about odors. This was put forward by the Dutch physiologist Hendrik Zwaardemaker (1857–1930) and emerged in 1876 under the influence of Otto Nikolaus Witt's "chromophore theory." According to this theory, the color of organic compounds is determined by functional groups known as chromophores. By analogy, in 1895 Zwaardemaker assumed that the smell of chemical compounds was also due to similar groups, which he called "osmophores" (from οσμή, osme, Gr.—smell, and φέρω, fero, Gr.—to carry). His hypothesis was therefore called "osmophoric." According to Zwaardemaker, whose theory was heavily criticized and rejected in 1920, osmophoric groups are -COOH, -COOR, -CHO, >C=O, -OH, and -NO_2. In the mid-twentieth century, new hypotheses were added, and by the end of the century there were more than 40. The large number of hypotheses regarding a scientific problem is, firstly, an indication of the public's interest in it and, secondly, of its complexity. Based on current experience, it is mainly organic compounds that have an odor, but inorganic substances such as the halogen elements fluorine, chlorine, bromine, and iodine also do. Ozone, phosphorus, arsenic, sulfur, selenium, and some of their compounds also have a smell.

I. G. Ivanov, *The Invisible Language of Nature*,
https://doi.org/10.1007/978-3-662-73302-8_15

Odor-intensive organic compounds can conventionally be divided into three groups: a) substances with similar structure and similar odor; b) substances with similar structure and different odor; c) substances with different structure and similar odor. The first group includes the aromatic hydrocarbons of the benzene group. The stereoisomers of unsaturated compounds belong to the second group (e.g., cis-3-hexenol smells like green leaves, and trans-3-hexenol like chrysanthemums). The third group comprises compounds with a bitter almond odor such as hydrocyanic acid, benzaldehyde, and nitrobenzene, which have comparable odors but completely different chemical structures. Such a classification can drive any chemist to despair and prompt them to abandon the search for a relationship between chemical structure and odor. For example, M. Beats writes: "When reviewing the publications on odor, one cannot help but notice not only the sheer volume of material and the numerous attempts to bring clarity to this difficult problem, but also the futility of many of these attempts." This inevitably raises the question: If there is no clear relationship between the chemical structure and the odor of substances, is it not possible that there is a relationship between odor and some other molecular parameter of these substances?

The British researcher John Earnest Amoore (1930–1998) believed that the odor of organic compounds is determined by the shape and size of their molecules and, in some cases, by their electronic structure. In 1952, while still a doctoral student at the California Institute of Technology, he formulated his stereochemical theory of odors. This was a further development of work on the existence of several types of primary odors by R. W. Moncrieff from the 1940s. The molecules of the substances that determine these odors interact with specialized receptors according to the lock-and-key principle. The latter, in turn, is borrowed from Emil Fischer's (1852–1919, Nobel Prize in Chemistry 1902) concept of the binding of enzymes to their substrates. Originally, Moncrieff assumed the existence of 6 primary odors, but then increased the number to 20–30.

Amoore constructed spatial models of 616 organic compounds from various molecular classes and determined the geometric parameters of their molecules. This led him to postulate the existence of 7 primary odors: ethereal, camphoraceous, musky, floral, minty, pungent, and putrid, with each of these odors corresponding to a specific size and shape of molecule. The ethereal odor, for example, is characteristic of substances with rod-shaped molecules with a diameter of 5 angstroms (1 Å = 0.1 μm); the camphoraceous odor is characteristic of hemispherical molecules with a diameter of 7.5 Å; the musky odor is characteristic of flat, disc-shaped molecules with a short end; mint—wedge-shaped molecules with a polar group at the tip;

pungent—electrophilic compounds of the type R-N=C=S, R-CH=CH-CHO, R-CHI-COOR; putrid—nucleophilic compounds of the type R-NH_2, R-SH, R-SeH, R-TeR, etc. According to Amoore's theory, 3 of the main odors (esters, camphor, musk) are determined by both the shape and size of the molecules; 2 (floral and mint) only by the shape, and the other two (pungent and putrid) by the electronic structure, independent of the geometric parameters of the molecule. All other odors are complex and result from the combination of the 7 basic odors. For example, according to Amoore, the odor of anise is due to a mixture of camphor and floral odors; the odor of cedar is a combination of camphor, musk, floral, and mint; the odor of lemon consists of camphor, floral, mint, and putrid; the odor of garlic of ester and putrid, and so on.

He assumes that the 7 basic odors correspond to 7 types of olfactory receptors. On their surface are indentations that match the shape and size of the molecules of the odorants. Substances whose odor can be assigned to one of the basic odors bind only to one type of receptor, while others (with complex odors) interact with more than one receptor.

Amoore's theory quickly gained great popularity. He himself stated that it was universal and comprehensive, writing in 1963: "The stereochemical theory is simple, there are no exceptions, and it can predict the odor of unknown compounds." And indeed, Johnson and Sandowall synthesized the previously unknown 4-methylcarbomethoxypimelate and its derivatives, whose odor was predicted by the stereochemical theory. The same theory also predicted the musky odor of some new nitro compounds. The stereochemical theory is supported by the disease anosmia. This is a kind of "smell blindness," which manifests itself in the inability of some people to perceive certain odors. According to statistics, about 10% of people cannot smell hydrocyanic acid, which is reminiscent of bitter almonds. And one in a thousand cannot smell the nauseating odor of a skunk. According to Amoore, each type of specific anosmia is due to defects in the receptors for one or more basic odors. From the types of anosmia, one can therefore infer the number of primary odors. Because of his work, Amoore became one of the best-known anosmia researchers of the last century.

A little more than a decade after the publication of the stereochemical theory of odors, the euphoria about it had faded. Numerous publications appeared that questioned it. One of the first "problem cases" was, of all things, the odor of hydrocyanic acid. According to Amoore, it is a complex odor composed of camphor, floral, and mint odors. However, the hydrocyanic acid molecule is so small that it cannot interact with 3 receptors at the same time. Another example is acetonitrile (CH_3CN) and methyl isocyanide

(CH_3NC). Both compounds have small molecules of the same shape and size and, according to the stereochemical theory, should have the same odor. In practice, however, the first compound has a pleasant ethereal odor, the second a strong pungent odor. The same applies to ethanol (C_2H_5OH) and mercaptoethanol (C_2H_5SH). The former has a pleasant odor, the latter a nauseating one. All this shook the foundations of the stereochemical theory. Moncrieff wrote in 1967, not without a touch of pessimism: "Completely unexpectedly, at the 1965 symposium, Amoore's theory was refuted .... Thus came the end of one of the many theories about odors. In retrospect, it is clear that there was never really convincing evidence for this theory. On the contrary, questions kept arising that it could not address. Nevertheless, this theory contains many useful ideas."

Amoore's stereochemical theory was replaced by the no less popular wave hypothesis of R. W. Wright. According to this, odor is not determined by the shape and size of molecules, but by the vibrational movements of their atoms. And if molecules with different structures smell the same, it is only because their vibrational spectra have similar characteristic frequencies. Wright assumes that the molecules of odorants, when colliding with oxygen and nitrogen molecules, enter an excited state and, upon returning to their ground state, release energy that is absorbed by the olfactory receptors. At a temperature of 30–35 °C (about the temperature of the air in the nose), the released electromagnetic energy is in the infrared range (400–50 cm^{-1}), which is why a correlation must be sought between the odor and infrared spectra of the compounds. In contrast to Amoore, Wright determined the number of specific osmoreceptors to be 25–30, which characterizes the olfactory apparatus as a system with enormous information capacity. According to the wave hypothesis, camphor and musk odors are no longer primary, since the vibrational spectra of compounds with such odors show several maxima in the vibrational range 400–50 cm^{-1}. Wright found that many compounds of different structure with a musky odor have the same resonance frequencies (at 90, 150, and 180 cm^{-1}). This raised another question: How is the wave energy transferred from the molecule to the receptor?

Wright hypothesized that the absorption of energy involves pigment substances of the carotenoid group, which are highly unsaturated and occur in the olfactory epithelium. The energy of the oscillatory movements of the atoms in the molecules of the odorants corresponds to the energy of their conversion from the trans to the cis form. In other words, the molecules of the odor pigments, under the influence of the odorant, change from the trans to the higher-energy cis state. This is the beginning of the processes that lead to depolarization of the olfactory receptor and the generation of a nerve impulse. It is the first hypothesis that attempts to explain the significance and function of pigment substances in the olfactory epithelium for odor reception. It is supported by the fact that all noses whose epithelium is depigmented suffer from smell blindness. In other words: according to Wright's concept, the olfactory epithelium is a kind of "visual purple" that perceives infrared rays, which are invisible to our eyes. Of course, Wright's hypothesis has many weaknesses, which is why it too was abandoned. For example, it is undisputed that an odor sensation only arises when the odorant is adsorbed onto the olfactory epithelium. This leads to characteristic frequencies, and the question arises as to which frequencies determine the primary odors. Those given by Wright for pure substances, or others characteristic of the substance-receptor complex? It is known that the various optical isomers of carvone (a component of essential oils) have the same infrared spectra but differ greatly in odor. There is also the reverse case—the fly sex pheromone p-hydroxyphenyl-2-butanone acetate retains its odor after deuteration (i.e., replacement of its hydrogen atoms by the heavier hydrogen isotope deuterium), although its infrared spectrum is significantly altered. Be that as it may, Wright's hypothesis brought an exotic element to odor research, but ultimately could not prevail.

Proponents of the adsorption theory fall into another extreme. They believe that the molecules of odorants are adsorbed onto the surface of the receptor and that the released heat of adsorption is somehow converted into a nerve impulse. And since different substances release different amounts of heat, they also have different odors. Proponents of the adsorption theory deny the significance of carotenoids in the olfactory receptors for odor perception and attribute greater importance to phospholipids. According to the electrophysiological theory, the olfactory receptors are constantly charged microcapacitors. When the molecule of an odorous substance approaches them, they are discharged. This leads to a nerve impulse. Any molecule can cause a discharge of the capacitor. It must have a certain shape and a certain dipole moment, which determines its odor.

A review of the most common theories about odor makes it clear that there is still not a single one that satisfactorily explains the relationship between the chemical structure and the odor of organic compounds. Each of them is based on a more or less broad experimental foundation and explains a certain range of phenomena, but none can capture the complex process of odor perception as a whole. Interestingly, the various theories do not contradict but rather complement each other, and each contributes something to our knowledge of the functions of the olfactory epithelium. This is a good prerequisite for the development of a more general theory that unites the rational evaluation of all previous hypotheses and provides a comprehensive overview of the relationship between chemical structure and the odor of substances, as well as the nature of the generation of the odor signal.

Why Is There Chaos in the Study of Odors?

Since odors are an attribute of organic compounds and their perception is a physiological act linked to the course of complex biochemical and biophysical processes, the science of olfaction is an interdisciplinary field at the intersection of chemistry, physiology, biochemistry, and biophysics. In the past and present centuries, these areas have been extremely productive and have yielded several Nobel Prizes. However, the establishment of a new interdisciplinary science takes time, and there are phases in which specialists from the "pure sciences" work in isolation and not in coordination with one another. For example, chemists working in the field of olfaction have little interest in receptor physiology. Physiologists, in turn, have little interest in the chemical structure of substances. As a result, findings are often obtained that do not fit together. This is similar to the anecdote of the three blind men and the elephant. For all three, it was an unknown animal. One day, they had the opportunity to touch it. One touched the ear, another the trunk, and the third the legs. When asked what the elephant was, the first replied: a large ear; the second: a large trumpet; and the third: a large foot. We cannot deny that each of them was right, but only a sighted person can understand that these are actually parts of one and the same body. Similarly, the organic chemist sees osmophoric groups in odorants, the stereochemist sees geometric figures, the biochemist sees substrates and reactions, the physical chemist sees electromagnetic waves, the electrophysiologist sees capacitors, and so on. It is obvious that the efforts of the individual specialists must be directed toward a common goal in order to grasp the process as a whole. In addition to the need for improved coordination among the various specialists,

I. G. Ivanov, *The Invisible Language of Nature*,
https://doi.org/10.1007/978-3-662-73302-8_16

there are also objective reasons for the lag in research in the field of olfaction. Above all, there is a lack of physical instruments for measuring odors. Odor is not a physical phenomenon, but is perceived subjectively. While in hearing and vision research the parameters of the stimulus (sound and electromagnetic waves) can be measured precisely and objectively, the effect of an odorant is reported subjectively. The same substance may be perceived by one person as pleasant-smelling, by another as unpleasant, and by yet another as odorless (in the case of anosmia). An example is the smell of gasoline. Some people cannot tolerate it, while others are so addicted to it that they need to inhale several milliliters of gasoline every day. The assessment of odor intensity is also subjective and depends on the acuity of the sense of smell. It does not matter whether the person smelling is a man or a woman. However, it is known that women have a more highly developed sense of smell than men. The time of day is also important in the study of odors, as olfactory acuity fluctuates circadianly. Generally, it is stronger in the morning and weaker in the evening.

We have already mentioned that dogs have a keener sense of smell than humans. A German shepherd was trained to recognize the smell of beryllium. It then became a search dog for beryllium-containing minerals. Because of this ability, dogs are often used as ore prospectors. The best of them can detect 20–26 different minerals, which means they can even smell simple chemical elements that we cannot detect.

Studies on odors have been conducted for a long time, but even today there are no unified concepts or established terminology in the science of olfaction. Terms such as “floral,” “musky,” “minty,” and others are too vague and flexible. They do not provide a solid scientific basis for in-depth research.

Another important reason for ambiguities in olfaction is the fact that chemists to this day do not know (or at least are not certain) how pure substances smell. It is generally known that hydrogen sulfide, pyridine, and skatole are symbols for compounds with disgusting odors. However, few people know that when fully purified, they are either odorless or even have a pleasant smell. For example, highly purified skatole smells like jasmine. In other words, the odor we attribute to a substance is often due to the impurities that accompany it. If an impurity is a constant companion of the substance and has a strong odor, it can mask or drastically alter the odor of the main substance. And what substance is truly pure? It turns out that there is no strict criterion here either. The purity of volatile organic compounds is checked by gas chromatography. But even with the most sensitive gas chromatographs, impurities with a content of less than 10^{-3} to $10^{-4\%}$ cannot be detected. However, if the chromatographic peaks of the impurity overlap with those of the main component, it cannot be detected at all. This does not apply to the sense of smell, however. For most organic compounds, the sensitivity of the human nose is 10 to 100 times higher than that of the most advanced gas chromatograph. Thus, if a substance is classified as "chemically pure," this does not mean that it is also "olfactorily pure."

Finally, there is another serious obstacle to the correct interpretation of the results of olfactory research. This is the variety of objects used in experiments. The relationship between odor and chemical structure is studied in humans. The physiological and biochemical processes involved in chemoreception are studied in animals. Due to the similarity of the anatomy of the olfactory organs in mammals, it is generally assumed that they function the same way in all animals. For this reason, the essential scientific data on the relationship between odor and chemical structure have been obtained in animals and apply to animal olfactory receptors. To what extent this also applies to humans is difficult to answer.

The Invisible Messages of Mammals

The language of smells is primarily an animal language, but even in humans it is part of their signaling system. In the previous chapter, we established that there is still no universally accepted classification of odors. One reason for this is that humans lack an abstract concept of smell. The olfaction specialist A. Bronstein states: "While we have the concepts of salty, sour, sweet, and bitter for taste—terms that describe properties of many substances—and for colors the concepts of blue, green, yellow, and other colors, which are characteristic of many objects, our ideas about smells are object-oriented. We cannot characterize a smell without naming the substance or object to which it refers." It is the concreteness of the information associated with smells that leads animals to trust their sense of smell. A dog may not recognize someone by voice or appearance and may start barking, but its behavior changes as soon as it smells them. The individual scent is the most reliable attribute for identifying its owner. That is why the human saying "I only trust my eyes!" in the language of dogs becomes "I only trust my nose!"

This also explains the different notions of familiarity between humans and dogs. Getting to know another dog means getting to know its scent. When dogs meet for the first time, they actively sniff each other for several minutes. There is a legend about the ritual of "sniffing" among dogs. Once, they sent a messenger to God to tell him about their hard canine life. Upon his return, the messenger was to be marked with a special scent. However, the messenger never returned, and since then the dogs have been searching for him.

I. G. Ivanov, *The Invisible Language of Nature*,
https://doi.org/10.1007/978-3-662-73302-8_17

For humans, whose sense of smell is of secondary importance, it is hard to imagine how much information animals receive through scents. Only people whose sense of smell has become secondarily heightened can understand this to a limited extent, as in the case of the American writer Helen Keller. She became blind and deaf at the age of one year and seven months. Since then, her world has consisted 100% of touch and smell. The same unfortunate fate befell the Russian writer Olga Ivanovna Skorokhodova, who writes in her famous book "How I Perceive the World Around Me": "I have become so accustomed to relying on my sense of smell that everything I perceive I seem also to see and hear."

It is well known that animals with a highly developed sense of smell are able to orient themselves in complete darkness, where sight is useless. A Bashkir hunter describes a case in which he tracked a moose for hours in the dense tundra. When he finally shot it, he discovered that it was blind in both eyes. Nevertheless, this did not prevent it from confidently locating forest paths and skillfully avoiding trees. Obviously, such animals perceive the scent of objects from a distance and orient themselves in their environment no worse than a sighted person. It is likely that objects have special olfactory dimensions for them, which they can "see" perfectly even in total darkness. If a blind animal has any chance of surviving in the wild, an animal without a sense of smell is doomed. The sense of smell is the only sensory function that cannot be compensated for by other senses—it is unique.

Smells serve not only for orientation but also contain other important information. An individual scent is a kind of letter through which its originator (predator, prey, or mate) can be uniquely identified. The scent left by an animal of the same species is particularly rich in information. It reveals the sex, age, size, strength, health status, mood, and many other characteristics of the animal that left the scent trail. Since the scent fades over time, the scent mark also contains information about when it was left. If, of two tracks, the newer one smells stronger, that is important information. In the language of animals, this means: "If it smells like an animal I can eat, I run in the direction of the stronger scent. If it is a predator, I run in the opposite direction."

The importance of the sense of smell for the lives of mammals is confirmed by the fact that most mammals are born blind and with underdeveloped hearing, but with a fully developed sense of smell. The example of the kangaroo is particularly striking. It gives birth to its young after a gestation period of 33 days. The kangaroo baby is essentially an embryo weighing only a few grams. After birth, it crawls to the mother's pouch, where it latches onto one of the four teats. It remains there for 235 days until its development is complete. For a long time, it was believed that the mother herself placed the newborn in the pouch. But it is now clear that the kangaroo female is not the best mother in the animal kingdom. During birth, she simply lies on her back or sits down and begins licking her pouch from the inside. After the young leaves the birth canal, it crawls like a worm into the mother's pouch. Since it is blind and deaf, it can hardly find its way in this "jungle" and can only orient itself by its sense of smell. It has been found that it is born with a fully developed sense of smell and that its olfactory bulb occupies more than half of its brain. It is not exactly known what attracts the baby to the pouch—the smell of milk or the mother's saliva. It is assumed that licking the inside of the pouch serves to guide the baby precisely there and to the milk glands inside.

The sense of smell is not equally developed in all mammals. Depending on the degree of development, they are classified as macrosmatics (very good sense of smell), microsmatics (poor sense of smell), and anosmatics (without a sense of smell). Macrosmatics include carnivores, rodents, ungulates, etc.; microsmatics are primates, humans, seals, and toothless whales; and anosmatics are dolphins. The acuity of the sense of smell in some macrosmatics exceeds our imagination. We have already mentioned the hypersensitive nose of the dog, which can detect a few molecules of butyric acid in the air. But the dog can also find deeply buried objects in the ground that are odorless to humans. Because of this unique ability, dogs are assigned important and responsible tasks. During World War II, they were used to detect mines. The Leningrad dog Dick, who discovered 11,720 mines, is legendary. Ore-detecting dogs, in turn, discovered ores lying 12 m underground. Avalanche dogs detect the scent of people buried under several meters of snow. In the cold forests of the north, squirrels have been observed finding pine cones under 1–2 m of snow. Dogs, wolves, foxes, beavers, and raccoons can detect the scent of humans at a distance of 100 m; bears, wild boars, roe deer, fallow deer, and others up to 500 m; reindeer, red deer, and moose up to 1000 m; and elephants from 1000–5000 m. According to some elephant hunters, it is practically impossible to approach a herd undetected in calm weather. It is only possible in windy weather, and then only if you approach from the leeward side. There is a curious case of a working elephant that refused to fill a pit with stones. When the reason was investigated, it was found that a small kitten was in the pit. The elephant only began to work after the kitten—and presumably its scent—was removed from the pit.

When comparing the olfactory acuity of macrosmatics, a correlation was found between the size of the animal and its olfactory sensitivity. This is explained on the one hand by the larger surface area of the olfactory epithelium and on the other by the greater volume of air inhaled, which introduces a larger amount of odorants. Large animals have another advantage. Due to their size, they are able to sniff the upper layers of air. When an elephant checks a scent, it holds its trunk 3–4 m above the ground. An interesting sight is a herd of elephants disturbed by approaching humans. They all look in one direction, with trunks and ears pointing toward the people. Animals with a highly developed sense of smell not only have a larger stature but also a moister nose. The nose of a cat, for example, is drier than that of a dog, a bull, a moose, etc. It is believed that a moist nose is necessary to determine wind direction and thus the direction of the scent. Hunters determine wind direction by moistening their finger and holding it up to the wind.

Musk Glands

Anyone who has ever had pets or visited a zoo knows that mammals smell bad—and that each species has its own distinctive odor. Even a microsmatic species like humans can easily distinguish the scent of a goat from that of a cow, a pig, a fox, and so on. There is an inverse correlation between the acuity of an animal's sense of smell and the strength of its own odor. Some of the most strongly smelling animals belong to the families Mephitidae (skunks) and Viverridae (civets). Rats and mice also have a strong odor.

The specific odor of mammals is determined by special glands called musk glands. These are exocrine glands (glands with external secretion) that are, in origin, a modification of sweat and sebaceous glands. They are specialized in producing odoriferous substances—musk—. The reader will probably notice that we use the term musk here instead of pheromone. Musk is an old term coined in medieval Europe to describe aromatic animal substances imported from India and China. However, the biological meaning of this term does not fully coincide with that of pheromone, since musk has functions other than those of pheromones. They serve not only reproduction, but also help maintain order within animal communities, mark paths and the boundaries of inhabited territory, mark their own young, and aid in wound healing. Many of their functions are probably still unknown.

Musk glands can be located in various places on the animal's body. In goats and antelopes, they are found on the head; in females, on the back; in camels, on the neck; in foxes, on the tail; in elephants, between the eyes and ears; in rabbits and cats, on the paws, and so on. In some cases, they have no direct opening to the body surface. Their secretion is released into the large

I. G. Ivanov, *The Invisible Language of Nature*,
https://doi.org/10.1007/978-3-662-73302-8_18

intestine and mixed with the excretions. They can also be located near the urinary tract, with their secretion then being passed into the urine. Musk glands also differ in their number. The saiga antelope (an ungulate from the cattle family), for example, has 24 different musk glands. The amount of musk secreted also varies. Animals whose glands produce a lot of secretion have an anatomically shaped sac (musk sac), which in the male capercaillie, for example, is located beneath the abdomen and holds 30–45 g of musk.

As a rule, the musk glands of male individuals are more developed and more numerous than those of females, which is why males have a stronger odor. In group-living animals, the activity of the musk glands depends on the animal's rank. Typically, the alpha animals smell the strongest. In other words: no one is allowed to smell more than the boss. While all wolves in a pack may seem the same to us, a new member of the pack can unmistakably recognize the rank of every other family member within minutes by scent. In this case, the musk glands serve the function of rank insignia in the army. If an individual becomes ill, the function of the musk glands is weakened and it loses its rank in the hierarchy. Thus, the musk glands also serve as an indicator of the health status of group members. This is not surprising, considering that exocrine glands are controlled by both the nervous system and the endocrine glands. Any disturbance in the activity of these two systems affects the function of the musk glands and leads to a change in the chemical composition of the musk. Therefore, the scent of musk is also a reflection of metabolism and the entire internal state of the animal.

Chemical Nature of Musk

In the study of the chemical composition of musk, researchers have shown a certain preference—animal musk, in which the perfume industry had a greater interest, was studied in more detail. One of the first types of musk to reach chemical laboratories was that of the Siberian musk deer (*Moschus moschiferus*), also known as the musk deer (Fig. 1). It is found in the Altai, Mongolia, China, Korea, Siberia, and on Sakhalin Island, and is a species of ruminant with a body length of up to 100 cm and a weight of about 15 kg. The males have a musk pouch on their abdomen, in which up to 45 g of a pleasantly scented, oily substance accumulates. This is highly valued in perfumery.

Because of the musk in its pouch, the Siberian musk deer was heavily poached in the past, but today it is a protected species. Musk consists of free fatty acids and phenols (10%), waxes (38%), and steroid compounds. The main component is a macrocyclic ketone (3-methylcyclopentadecanone) called muscone. The pleasant scent of musk inspired chemists to develop many analogues. Their research shows that the scent strongly depends on the size of the carbon ring. Compounds with 14–18 carbon rings have a pleasant odor, those with 10–12 smell like camphor, and those with 13 carbon rings smell like cedar. Compounds whose ring contains more than 18 carbon atoms are odorless.

Another macrocyclic ketone, civetone, has a musky scent. It is the main component of the musk of the Indian civet (*Viverra zibetha*, Fig. 2), which is widespread in India, Burma, China, Thailand, and Malaysia. It belongs to the family of civets (Viverridae), weighs 12 kg, and has a body length

I. G. Ivanov, *The Invisible Language of Nature*,
https://doi.org/10.1007/978-3-662-73302-8_19

Fig. 1 Siberian musk deer (*Moschus moschiferus*) and the main component of its musk: 3-methylcyclopentadecanone (muscone). (Photo: © Marvin Samuel Tolentino Pineda/Getty Images/iStock)

Fig. 2 Indian civet (*Viverra zibetha*) and the main component of its musk, cis-9-cycloheptadecen-1-one (civetone). (Photo: © karenfoleyphotography/Getty Images/iStock)

of 80 cm. Its color is gray with numerous dark stripes and spots. Its musk glands are located in the genital area and are extremely productive.

A civet can secrete up to 50 kg of musk annually. In addition to civetone, other macrocyclic alcohols and ketones are also produced. Civet musk is probably the oldest substance coveted by humanity. In his diary from January 1552, Antonio Pigafetta wrote: "Musk comes from China. It is obtained from the muskrat, which feeds on a shrub called 'Shamaru'. To obtain the musk, they apply leeches to the animals, let them drink the blood, and then crush them. The blood is collected in a bowl and dried in the sun for 4–5 days. Only then does the musk become a remedy. Anyone who owns such an animal is obliged to pay tribute to the emperor. The musk grains sent to Europe are pieces of young goat meat soaked in musk and dried. The natives call the animal 'Rolle' and the leech 'Fussel'." Civetone is found together with another macrocyclic ketone (exaltone, cyclopentadecanone) in the musk of the muskrat (*Ondatra zibethicus,* Fig. 3) and in the secretions of the Muscovy duck (*Cairina moschata,* Fig. 4), the musk turtle, and in alligators. The original range of the muskrat is North America. From there, it was introduced to Europe. From Russia, it reached

Fig. 3 Muskrat (*Ondatra zibethicus*). (© Grigorii_Pisotckii/Getty Images/iStock)

Fig. 4 Muscovy duck (*Cairina moschata*). (© Ralf Blechschmidt/Getty Images/iStock)

the Srebarna Biosphere Reserve near Silistra in Bulgaria and also migrated independently from Serbia via Lake Srebarna. It weighs up to 1 kg and is one of the most highly prized fur-bearing animals.

Among the pleasantly scented animal products, two more should be mentioned—spermaceti (cetaceum or white ambergris) and ambergris from the sperm whale (*Physeter catodon*). This member of the toothed whale order is a living perfume factory. Its large head, which makes up about one third of its 30-ton body, contains over 2 tons of spermaceti. This is a whitish, putty-like substance with a pleasant scent, which is also valued in the perfume industry. The name is derived from "sperma" and "cetus" (Latin for whale), as whalers once believed it to be the sperm of the sperm whale. The physiological significance of spermaceti is still unclear. One hypothesis suggests that it serves to keep the sperm whale's head above water, as it has a lower relative weight than water. Another suggests that it is an antiseptic agent that helps disinfect and heal injuries to the mouth more quickly. It is also suspected that it acts as a resonator for the sperm whale's ultrasonic echolocation.

Ambergris is a waste product from the digestive tract of sperm whales. According to the author of "Moby Dick," Herman Melville, ambergris "resembles a fatty, sticky, and extremely foul-smelling old cheese." This waxy substance is found in the large intestine and rectum of the whale and is sometimes excreted with the feces. Because it is lighter than water, ambergris floats on the surface. Immediately after excretion, it smells earthy. When stored in a closed container, it then develops a scent reminiscent of musk and jasmine. From the last century to the present, ambergris has been an indispensable component of the most expensive perfumes, which is why it was traded for many years at the price of gold. In 1953, whalers discovered a huge piece of ambergris weighing 413 kg on the beaches of Australia and sold it for $120,000. Today, the price of ambergris is much lower, as it is no longer collected from the water's surface but obtained from killed whales. According to reports from whaling companies, ambergris is found in 4–5% of killed whales. In addition to perfume production, ambergris is also used in folk medicine to treat epilepsy, rabies, heart disease, etc. The biological significance of ambergris is still unknown. Some scientists believe it is a pathological product of the gallbladder in sick whales, while others think it is a normal secretion of the rectal glands. Still others consider ambergris to be a protective substance that shields the whale's intestine from the sharp chitin remnants of consumed cephalopods. It may also be musk associated with sexual activity, as it is found only in males. Animals with a pleasant scent are an exception in the animal kingdom. Most of them emit an odor that is unpleasant to humans.

With few exceptions, deer do not have musk pouches. They secrete their scent diffusely over a large area of the body, which in turn makes it difficult to study its chemical composition. The main component of the musk of the American mule deer (*Odocoileus hemionus*, Fig. 5) is thought to be 4-hydroxydodecenoic acid, which can also occur as a lactone. In contrast, the anal gland of the pronghorn (*Antilocapra americana*, Fig. 6) secretes a mixture of valeric and isovaleric acids as well as their esters with aliphatic alcohols and branched carbon chains. Forty-five different chemical compounds have been found in beaver musk, and the secretion of the mouse's musk glands is also multi-component.

From the sparse literature data known so far, it can be concluded that the musk of most animals consists of complex mixtures of volatile carboxylic acids with 2–6 carbon atoms and their esters with aliphatic alcohols. It is assumed that the chemical composition of animal secretions is

Fig. 5 American mule deer (*Odocoileus hemionus*). (© Pedro Carrilho/Getty Images/iStock)

Fig. 6 Pronghorn (*Antilocapra americana*). (© John Morrison/Getty Images/iStock)

species-specific and that the quantitative ratio of the various components is individual. In general, the qualitative and quantitative composition of musk is genetically determined and reflects the uniqueness of the individual. By varying the concentration of the components in the chemical mixture, thousands of nuances of the same scent can be produced. This allows animals of the same species to have both a similar species-specific and a different individual scent, characteristic for each member of the group. Only identical twins smell the same.

The olfactory apparatus of each individual is tuned to its own scent, with which it compares all other individuals. In this case, the individual scent acts as a mirror through which the animal compares its image with that of its conspecifics. Here, we can compare the functions of the olfactory apparatus to tuning a radio receiver. When the animal smells an object, it first determines whether the scent it perceives comes from an animal. If it is an animal, the scent is automatically compared with its own to determine whether it belongs to an animal of the same species or not. To continue the analogy with the radio receiver, one could say that a narrower frequency range is selected within the already chosen range. If it turns out to be the scent of a conspecific, a further step of refinement follows—whether it is identical to its own scent or not. If it is the scent of another individual, it must be further analyzed to determine sex, rank, health, etc. This corresponds to the fine-tuning of the radio receiver when selecting a specific radio station.

The musk glands of newborn animals are underdeveloped. The pouch of the young capercaillie is wrinkled and empty and only begins to fill with musk in the third year of life. The musk glands of young animals, foxes, and rodents begin to function 2–3 weeks after birth. At this stage, they are practically odorless and carry only the scent of the mother, which in turn helps her find them more quickly. Overcoming the scent barrier is the most difficult phase when transferring young animals to another litter. Konrad Lorenz describes the reaction of the nursing dog Senta to overcoming the scent barrier after he brought a small, homeless dingo into her litter: "She immediately bent over the screaming baby, mouth wide open, ready to pick it up and carry it to her puppies. And at that moment she was struck by the foreign scent the dingo brought from the zoo. She recoiled in shock and inhaled the air with a whistling sound, such as I had never heard before or since from a dog. I took not only the dingo but also all of Senta's puppies, put them in a basket near the kitchen stove, and left them there overnight so they could rub against each other and mix their scents. When I brought the puppies to Senta the next morning, she received them with some suspicion and became very agitated, but soon carried them into her kennel, the little dingo along with them."

Once the scent barrier is overcome, nursing dogs and animal mothers tend to feed anyone who is hungry. There have been cases where a mare nursed a calf, a cow a pig, a dog a pig and/or a deer, a cat a rat, and so on.

Fragrant Calling Cards

Humans and animals have different ideas of comfort. While humans prefer a pleasant environment free from intrusive odors, animals feel comfortable in surroundings marked by their own scent. This leads to a conflict between humans and animals. To make their homes cozy, people clean and air out their apartments, which can be unpleasant for their pets. Conversely, pets create problems for their human housemates through their efforts to make the environment comfortable for themselves.

The first concern of any animal when settling into a new home is to explore it and leave its fragrant "calling cards." When a mongoose (Herpestidae), a family of mammals in the order Carnivora and mortal enemy of the cobra, enters a room, it first sniffs all the objects and then begins to mark them with its musk. The process is as follows: it sniffs the object, turns its back to it, stands on its front paws, lifts its tail, and presses its hindquarters (where the musk glands are located) firmly against the object. In this way, it marks the floor and all the furniture in the room. If the scent fades, the mongoose reinforces it. The animals also use musk to mark the boundaries of their territories. These are the areas where they feed and mate. There are no visible boundaries between the different territories, but there are "scent marks"—a kind of coat of arms of the ruler. Gerald Durrell describes the marking of the territory of one of his raccoons as follows: "I tied the two animals to the trees with long ribbons. Each time, Matthias spent ten minutes marking his territory with the foul-smelling secretion from the gland at the base of his tail. He would solemnly walk in a circle, with an expression of utmost concentration, and from time to time

I. G. Ivanov, *The Invisible Language of Nature*,
https://doi.org/10.1007/978-3-662-73302-8_20

squat down to rub his backside against a stone or stick. With this ritual, which is equivalent to raising a flag over conquered land, he became calmer and, with a clear conscience, began to catch beetles."

Deer, roe deer, and antelopes also mark their territory. Since their musk glands are located on their heads, they rub their heads against trees. Occasionally, one can even observe a "mock fight" with a tree, the purpose of which is to leave a strong scent mark. The distance between these "boundary posts" is about 5 m. Cats, which actively mark their territory in March, stand next to the chosen object, stroke it, arch their backs, sit on their hind legs, and nervously raise their tails, after which a drop of fluid is released from the anal musk glands. Animals whose musk glands are connected to the urinary tract or rectum mark objects in their territory with urine or excrement. Anyone who has ever walked a dog has noticed that it constantly sniffs as if searching for something, stops at certain spots, lifts its leg, and releases a little urine. Konrad Lorenz describes this behavior as follows: "The lifting of the dog's leg has a very specific meaning. As paradoxical as it may sound, it is exactly the same as the song of the nightingale. It is a way of marking the boundaries of one's own territory and warning others not to enter. The well-trained dog refrains from this marking in its own home, as the air there is already saturated with its scent. But if a strange dog or even a sworn enemy crosses the threshold of the house for even a second, the learned manners disappear instantly and the natural instinct comes out in full force. Every dog with any character at all then considers it his sacred duty to destroy the unfriendly scent of the stranger and to leave his own, even stronger scent mark. To the horror of its owner, the otherwise well-behaved pet makes a round through the room and shamelessly lifts its leg at every chair, table, and cupboard. So think carefully before you visit the home of a dog owner who has befriended your dog!"

Everywhere on the streets and in gardens where dogs move, there are "calling cards." They sniff along the paths and search for them. Every time your dog finds one, he leaves his own. And he adjusts the amount of urine according to the physical characteristics of his fellow dog. The rule is simple: more urine is needed to overpower the scent of a larger rival, and conversely, less if the rival is smaller. It has been observed that animals mark a place more intensely when in a state of emotional excitement. For example, a confined fox begins to smell more strongly when a wolf passes by. When a cat or rabbit encounters a dog, the undersides of their paws become moist with secretions from the mucous glands located there. When a dog or wolf is ready to fight, it raises its tail high so that the scent from the perianal musk glands can spread more effectively.

The Biological Significance of Fragrant Calling Cards

Marking with musk primarily serves orientation and exploration, helping the animal to recognize its habitat more quickly. The animal's behavior is governed by instincts. Through unconditioned reflexes, it feeds, drinks water, and reproduces. However, the successful fulfillment of its tasks for the species as well as its own well-being depends on its ability to learn about its environment. In every new environment, an animal adapts to the specific living conditions. Therefore, its first task is to become thoroughly familiar with its habitat. Acclimatization consists of sniffing and marking the surrounding objects, which in the animal's language means: "Check! There is nothing dangerous here and I might be able to stay." If location recognition were to occur through visual perception, more time would have to be spent to create a lasting conditioned reflex. Leaving a scent helps the animal quickly gain life experience, avoid dangers, and conserve nervous energy. A territory marked with its own scent creates confidence, and the "calling cards" left behind are a stress factor for an intruder. Cases have been observed in which a weaker animal in its own territory repels the attack of a much stronger opponent because the latter was already deterred by the owner's preventive scent marks.

"Calling cards" serve another purpose as well. During the breeding season, they are directed at the fairer sex. Each male fences off his territory, in which he waits for his chosen mate. Here, competition is fierce, especially among polygamous animals, and the scent of the territory is decisive. Since territories with a stronger scent are preferred, the males compete to perfume them lavishly. The logic is simple—more musk is secreted by the sexually

I. G. Ivanov, *The Invisible Language of Nature*,
https://doi.org/10.1007/978-3-662-73302-8_21

more active males. For this reason, it is common during the breeding season for several females to gather around one male, while others are ignored. This apparent injustice, however, is justified from the perspective of natural selection—sexually active males are physically healthier and leave behind stronger offspring.

Musk and Demography

During the mating season, the secretions of the musk glands serve as sex pheromones. They not only attract individuals of the opposite sex but also stimulate and drive them to mate. There is ample evidence that the musk glands are closely related to the gonads. First: their activity is cyclical, peaking during the mating season. At this time, they are large and filled with abundant secretion, which is why the animals have their strongest odor during this period. Second: castration suppresses the activity of the musk glands. Castrated animals have a weaker odor and do not attract the opposite sex. However, if sex hormones are injected, the function of the musk glands is restored and the animals become attractive to the opposite sex again.

The chemical nature of true sex pheromones in higher animals has also been elucidated in individual cases. For example, it has been found that the function of the sex pheromone in wild boar (Fig. 1) is performed by the steroid compound 5-alpha-androst-16-en-3-one.

It is secreted by the submandibular glands of the male and attracts the female during the period of receptivity. Under its influence, she literally loses her mobility and allows the male to approach her. In contrast, the sex pheromones of monkeys are a mixture of acetic, propionic, butyric, isomaleic, and isovaleric acids. During the mating season, the urine of females in almost all animals acts as an attractant for males. If a non-receptive female dog is sprayed with the urine of a receptive one, the males begin to court her and attempt to copulate. This urine retains its stimulating effect for up to one minute. The same effect has been observed in mice.

I. G. Ivanov, *The Invisible Language of Nature*,
https://doi.org/10.1007/978-3-662-73302-8_22

Fig. 1 Wild boar (*Sus scrofa*) and its male sex pheromone 5-alpha-androst-16-en-3-one. (Photo: © ViktorCap/Getty Images/iStock)

In mice, a sex pheromone is secreted that regulates population density. It is believed that this is a general property of sex pheromones in mammals. This is not observed in lower organisms. It has long been known that a colony of rats or mice, kept in a confined space and supplied with food and water, does not increase in number without limit. It stabilizes at a level that is maintained over a long period without an increase in mortality. The population usually stabilizes when the number of animals per square meter exceeds that of their natural habitat. This is due to pheromones that inhibit the reproductive process and stabilize the population. If the living space is enlarged, the release of these pheromones decreases and the group size increases until the next saturation limit is reached. What is the mechanism of action of these pheromones?

Most studies have been conducted with mice. It has been found that females are extremely sensitive not only to male sex pheromones but also to the pheromones of their sisters. For example, if two females are placed in a nest in the absence of a male, they mutually influence the length of their estrous cycle. While the cycle under natural conditions lasts 4–5 days, in the absence of a male it is extended to 11–12 days. There have even been cases of pseudopregnancy, i.e., a complete suppression of the sexual cycle. This phenomenon, known as the Lee-Boot effect (synchronization of the menstrual cycles of female mice), is more pronounced the higher the population density and the smaller the living space. The strong odor of female mice, even in the absence of the animals themselves, is sufficient to cause hypertrophy of the adrenal glands, leading to a reduction in reproductive capacity (Ropartz effect). If a male is introduced into a group of female mice with suppressed cycles, the cycle is restored and, at the same time, the sexual cycles of the individual mice are synchronized. The same effect is observed if, instead of a living male, only the urine of a sexually mature male is introduced into the group. This phenomenon is known as the Whitten effect. The activity of male urine is more than surprising. In wild house mice, the Whitten effect has been observed with the addition of only 0.01–0.1 ml of urine. Attempts have been made to determine the chemical nature of the active substance causing these effects. However, the results are inconclusive. Since it is assumed that sex pheromones are produced in the testes, the effect of aqueous extracts from mouse testes was investigated. However, these have proven to be inactive. It is, however, known with certainty that the substances causing the Whitten effect are species-specific. Another effect related to the action of sex pheromones, the Bruce effect, has been observed in mice. If a pregnant mouse is exposed to male urine from another group, the pregnancy can be interrupted.

The phenomena described illustrate the great ecological significance of sex pheromones in mammals. While under normal conditions they stimulate sexual activity and facilitate reproduction, in emergency situations (e.g., when the living space cannot be expanded) they act as a regulator of population density.

A Bit More About Musk

Animal musk is also attributed with a protective function. As early as 1873, Darwin wrote: "Both sexes of ground squirrels are equipped with strongly scented abdominal glands. Considering that their carcasses remain untouched by birds and predators, one should not doubt that their scent protects them."

Since ancient times, animal musk has been used in folk medicine by many peoples as a remedy. For example, more than 50 medicines for the treatment of diseases, including skin conditions, were made from beaver musk, the so-called castoreum. Recent research has shown that the healing effect is due to the high salicylate content. For this reason, and because of their valuable fur, beavers have been hunted en masse for centuries and are nearly extinct. In 1935, the Russian zoologist Anatolij Fedjuschin wrote: "It can be said with good reason that castoreum has destroyed the beaver." Today, the beaver is protected in almost all countries.

The musk of the muskrat also has a healing effect. Both castoreum and muskrat musk have bactericidal properties and are used to treat urinary tract infections. Many types of animal musk have a plant-like odor. For example, musk from animals of the ferret family smells like a mixture of garlic and onion with nuances. Beaver musk, on the other hand, has a phenolic odor, and muskrat musk smells like a blend of mint and anise.

Most types of musk are insect-repellent; mink and raccoon musk repel mosquitoes. The secretions from the glands on the paws of animals from the cat family also have a repellent effect. Joy Adamson, author of the popular book "Born Free," writes: "Twice I saw Elsa unhesitatingly cross the broad

I. G. Ivanov, *The Invisible Language of Nature*,
https://doi.org/10.1007/978-3-662-73302-8_23

stream of black ants, causing turmoil in their ranks with her large paws. These little warriors attacked any obstacle in their path, but they did not touch Elsa." It is possible that the musk secreted from the paws of lions and tigers serves as a natural protection against scorpions and other poisonous insects, which are abundant in tropical countries. The products of the ear glands, as well as earwax itself, also have an insect-repellent effect. As a result, practically no insects enter the ever-open ear.

It is known that animals are not insensitive to sedatives and anesthetics. The interest of elephants in alcoholic beverages has long been known. In South Africa, the marula tree grows, whose fruits, about the size of tangerines, contain a lot of sugar. Once ripe and fallen to the ground, they begin to ferment and turn into small alcohol-containing (about 3%) balls. At this point, they are especially attractive to elephants. They eat them, which leads to intoxication. Time and again, drunken and rampaging elephants have been observed attacking people and villages. This weakness is exploited in zoos and circuses to administer medicine previously dissolved in an alcoholic beverage. Monkeys also enjoy drinking. Other animals eat juniper and juniper berries, which have a calming effect. A classic case of drug dependence in animals is the fondness of cats for valerian. Incidentally, this weakness is characteristic of all members of the cat family. For them, valerian is a drug. A cat was observed systematically consuming valerian for years. Gradually, it neglected its hygiene, its fur became dry and matted, and it took on the appearance of a drug addict. The attraction of valerian for cats usually manifests in adulthood and is more pronounced in males. If a cat eats valerian roots, it falls into a narcotic sleep. At very high doses, death can occur. Studies on the physiological effect of the plant on cats have shown that it is also a sexual stimulant, i.e., the intake of valerian has sexual overtones. This is not surprising, since the sex pheromones of higher mammals are a mixture of lower fatty acids and their esters, among which the esters of valeric acid predominate. For information, it should be noted that the main component of valerian essential oil is the borneol ester of isovaleric acid.

In One's Own Skin with a Foreign Scent

In his short story "Domino," Ernest Thompson Seton describes the reaction of a silver fox upon discovering a piece of meat left as bait: "He was fascinated. The strange movements of his body showed that he was losing control of himself. He tried to get as close as possible, to intoxicate himself with the scent, to absorb its essence. He would have liked nothing better than to be enveloped by it. In anticipation of pleasure, he turned his head to the side, pressed his neck against the contaminated ground, and bristled his fur. He rolled onto his back and growled."

Rolling on food or the dead body of prey is a common reaction among predators. This is referred to as the tergore reaction or tergore reflex (from the Latin words "tergo"—backside and "tergoro"—to cover). Its purpose is to thoroughly coat the body with the scent of the prey. The ritual varies among different animals. While wolves, jackals, foxes, etc., roll themselves, members of the cat family rub the foul-smelling substance into their fur with their forepaws. The bear, on the other hand, sits on the ground, grabs the smelly object with its paws, and rubs itself with it as if with soap, starting at the neck and head and then the shoulders. In the process, it sniffs, pants, shakes its head, and drools from its mouth. Its gaze is as focused as if performing an extremely important task.

The ability of hundreds of different odors and chemicals to trigger a tergogenic reaction has been tested experimentally. It was found that the smell of decaying meat has the strongest effect. What is the biological significance of the tergogenic reaction?

I. G. Ivanov, *The Invisible Language of Nature*,
https://doi.org/10.1007/978-3-662-73302-8_24

Like all innate instincts, the tergore reaction is beneficial for species living under natural conditions. One of its most important functions is signaling and information. By joining a group, the "fragrant" animal brings a "sample" of the delicacy and thus stimulates the appetite of the others. After sniffing the messenger, they follow its trail and quickly find the prey. However, this reaction can also serve to "deceive the enemy," i.e., it fulfills the function of olfactory mimicry. As we have already seen, not only predators are macrosmatic, but so are their prey. They perceive the scent of the predator from afar and look for a way to escape. Therefore, obtaining food is no easy task for predators. However, if the predator manages to mask its own scent, the situation changes. Of course, this introduces another problem. If a predator rubs its body with the scent of a herbivorous animal, another predator might become interested in it. Therefore, when a predator undertakes this risky procedure, it must have confidence in its own strength. Observations have shown that this reaction occurs more frequently in large and strong predators such as lions, tigers, lynxes, bears, and wolves, and relatively less often in smaller animals. Olfactory mimicry is also practiced by humans during hunting. For example, elephant hunters smear their bodies and shoes with elephant dung, and fox hunters moisten their hands and shoes with fox urine.

The Language of Our Ancestors

Since Aristotle, it has been assumed that humans occupy the highest position when living beings are arranged in ascending order of their complexity and perfection. This is the so-called scala naturae, which, however, only holds true if the ranking is based on the complexity of the nervous system. With regard to other organs and systems, this order cannot be maintained. For example, if one takes the sensitivity of the olfactory system as a criterion, then "… modern humans seem to be a degenerate animal for whom the language of smells has long since lost its original deep meaning" (according to the French scientist M. Barbier). And yet, there is ample evidence that, while chemical signals are not vital for modern humans, they were essential for primitive humans. According to some experts, the chemical mode of communication in *Homo sapiens* is still not without significance today.

Every person has their own unique, individual scent. Dogs know this best, as they can unmistakably recognize their ownerφs scent among thousands of other people. The scent of humans, as with mammals in general, is genetically determined. Only identical twins smell the same. There is a famous case in which twin brothers, reunited after 33 years of separation, could not be distinguished by the family dog. Racial, group, and individual scents are also perceptible to humans. The French writer Joris-Karl Huysmans writes in his short story "Against the Grain" about the scent of Parisian girls: "It is pungent and sometimes bothersome in brunettes, sharp and strong in redheads, intoxicating and penetrating in blondes. One could say it matches their way of kissing."

I. G. Ivanov, *The Invisible Language of Nature*,
https://doi.org/10.1007/978-3-662-73302-8_25

From the time when humans trusted their noses more than their eyes, some peoples have likely retained the custom of sniffing each other upon meeting. For example, people in southeastern India press their nose to the guest's cheek and inhale deeply. In fact, they do not kiss, but sniff each other. A similar custom exists among the Maori, the indigenous people of New Zealand. When they meet, they touch noses and remain in this pose for several seconds. Similar rituals are also observed among Polynesians, Malays, and other peoples. Presumably, this is a remnant from ancient times when humans identified and accepted strangers by their scent. This ritual is similar to our custom of kissing when greeting guests. Among men, it consists only of touching cheeks.

What determines a person's scent? The human body is equipped with numerous sweat glands, whose main function is thermoregulation. However, along with sweat, they also secrete complex mixtures of lower fatty acids (propionic, butyric, valeric, caproic acid, etc.) and their esters, which are responsible for the smell of sweat. The sweat glands of the feet are numerous, reaching up to $1000/cm^2$. They are not particularly important for thermoregulation, but are quite active and occasionally cause aesthetic problems. However, they once served to mark territory and paths to food sources, just as animals still do today. In addition to these sweat glands, the human body also has those considered musk glands. For example, these are the sweat glands in the anal and genital regions, the glands under the armpits, and on the hairy areas of the male chest. In childhood, they have no function, but become active upon reaching sexual maturity. This activity is therefore considered to be associated with sexual processes. It is likely that their secretions once acted as sexual pheromones, and that the abundant hair in their vicinity helped to better disperse them. A steroidal alcohol, identical to the sexual pheromone in boars, has been isolated from the secretions of the sweat glands of the male chest.

The musk glands are also responsible for the different scent of male and female bodies. Research indicates that human scent is determined by three main components, one of which smells like musk, the second is trimethylamine (fishy smelling) and the third is referred to as the alkali factor. In men, the alkali factor is 1,5-diaminopentane, also known as cadaverine; in women, it is coumarin, which smells like freshly cut grass.

In a study of the vaginal secretions of 682 women, French scientists identified a mixture of five lower fatty acids: acetic acid, propionic acid, isobutyric acid, butyric acid, and isovaleric acid. These also function as female sexual pheromones in primates. They exert a strong attraction on males. It is believed that in the distant past, the "acidic" vaginal discharge had a similar effect on men. The assumption that this secretion also has a pheromone-like function in modern men is supported by the fact that the content of volatile acids in the vagina of women changes cyclically. In the first phase of the menstrual cycle, it increases, in the second it decreases. In women who use contraceptives, this cyclicity is absent.

There is evidence confirming the importance of pheromones for the sexual life of early and perhaps also modern humans. Women have been shown to be sensitive to the scent of a macrocyclic lactone called exaltolide, which acts as a male sexual pheromone in mammals. Female sensitivity is periodic, with a maximum coinciding with ovulation. Such cyclicity has not been observed in women taking contraceptives. However, the injection of estrogens increases sensitivity to it, just as estrogens increase the sensitivity of the male sense of smell to this compound.

The connection between nose—olfactory bulb—hypothalamus—sexual organs has been demonstrated in many animals, but is not outside the realm of reality in humans either. Neuroscientists believe that the fitness of the sense of smell is important for the psyche and human health. It has been observed that in both sexes, a congenital underdevelopment of the olfactory bulb is associated with infantilism of the sexual organs. Some authors believe that, in psychosexual terms, humans are not inferior macrosmatics to dogs. However, they consciously suppress the role of their sense of smell. M. M. Gradek writes: "Even the most cultivated person sometimes lets their nose decide in delicate matters of love." The importance of the sense of smell for human sexual life is illustrated by another interesting effect, similar to the Whitten effect in mice. It consists of the unconscious synchronization of the menstrual cycle among young women living in dormitories.

The human sense of smell develops very early. Barely a year old, children already accept or reject food based on its smell. When sleeping infants are brought to the breast of nursing women, it has been observed that two-day-old babies do not recognize their mother's scent; two-week-old babies prefer their biological mother's breast, but their reactions are still indecisive; six-week-old babies show a very clear preference for their mother's breast. The scent that attracts them is not yet known. It is also not yet certain to what extent a person's individual scent is due to exocrine secretory glands and to what extent it is influenced by the activity of bacteria, the microbiome, which colonize the sweat and sebaceous glands of the skin. Some of these bacteria are symbiotic for humans, and their activity depends on the person's physiological and health status. Since we have repeatedly established that the sense of smell is of secondary importance for human life, the reader will probably ask: "Why is it necessary, then, to study chemical signaling in humans?"

Scientists believe that the advances in this field are not negligible and can be beneficial. It has long been known that a number of diseases are accompanied by a characteristic change in body odor. A precise analysis of human body odor would make it possible to detect pathological components in skin secretions, which could serve for diagnosis. In this context, attempts have been made to develop an "electronic nose" for medical purposes, capable of detecting and analyzing volatile substances emitted by the human body. How does it work? The patient is placed in a special chamber and surrounded by an airflow that carries the volatile substances to an adsorber with the properties of a capacitor. The adsorbed substances change the contact potential of the capacitor plates, and the resulting electrical signals are amplified and transmitted to a recording device. Such devices would be useful for diagnosing a number of diseases, especially psychiatric disorders, for which there are often no objective criteria for assessing the patient's condition.

The analysis of an individual's scent can also be useful in forensic science. The scent left behind by a criminal has long been used as physical evidence for identification. However, if the criminal is caught too late, the scent dissipates and a sniffer dog cannot detect it. Nevertheless, since the criminal almost always leaves scent traces, whether on objects he has touched or in the air of the room he occupied, this scent could be analyzed at the time the crime is discovered, the data recorded, and later used when the criminal is apprehended. A device for an "electronic police dog" has been patented in the USA. The device is kept secret, but probably relies on gas chromatographic or mass spectrometric analysis of volatile human body secretions. Such a device would make it possible to create a "file" of the scents of the population, or at least of individuals in a criminal circle, similar to modern DNA databases or fingerprinting (dactyloscopy).

Knowledge of the chemical composition of the substances that make the human body smell is also useful for perfumery and cosmetics. Studies of chemoreception in animals have shown that their olfactory apparatus is most sensitive to the scent of their own secretions. Knowing the chemical composition of these secretions in humans would make it possible to create attractively scented perfumes, deodorants, and other cosmetics in which the pleasant components of the natural scent of the human body are recognizable, while the unpleasant ones are "silenced."

Finally, there are ideas in medicine to use the sense of smell for therapeutic purposes. As already mentioned, in all vertebrates, including humans, there is a natural physiological axis: nose—olfactory bulb—hypothalamus. And the hypothalamus is closely connected to the pituitary gland, which in turn controls the activity of all endocrine glands. Consequently, it is theoretically possible to influence the pituitary gland, and thus the other endocrine glands, through an appropriate chemical stimulus. Since an odorant acts on the sense of smell in very small amounts, the new "medications" would have two major advantages over classical ones: they would be inexpensive and likely free of side effects. One of the specialists in human olfaction, A. Komfort, writes: "Perhaps we will witness olfactory therapy as the crowning achievement of medicine."

The significance of the sense of smell and human body secretions for human behavior itself has not yet been fully investigated. It is possible that humans are causing themselves previously unimagined harm through their activities. Since natural body odor is normal and has accompanied humans throughout their long evolution, some scientists consider the currently fashionable removal of body hair and the excessive use of soaps, perfumes, and deodorants as the amputation of an important information channel. A. Komfort says: "No biologist today can say whether these amputation procedures will have any undesirable effects on future human behavior. The notion that organs such as the appendix and tonsils serve no purpose and can be removed without consequences is naive and belongs to the 19th century. I would add that the negative consequences of the excessive use of inadequately researched and uncontrolled new cosmetic products, as well as modern aesthetic medicine and dermatology, on the behavior of modern humans are already being felt." Neither A. Komfort nor we are advocating a literal return to nature or, for example, the abandonment of the achievements of modern cosmetics. Rather, we call for moderation and a reasonable attitude toward the products of civilization, until their harmlessness to human health and behavior has been investigated and scientifically proven.

A Dialogue Between Two Kingdoms: "Yes" and "No" in the Language of Chemistry

Paleontology teaches us that several hundred million years ago, our planet was covered with lush vegetation consisting of giant ferns, horsetails, and gymnosperms (naked seed plants). With the exception of a few species, these have since disappeared. In their place, the angiosperms (flowering plants) have emerged. Today, they make up 85% of the plant kingdom. During the era of the Earth's primordial flora, insects also appeared. With their enormous size, they adapted to the plants of that time. Since they had few natural enemies, they reproduced undisturbed to an incredible extent. Then, several hundred million years ago, a confrontation occurred between the two great kingdoms of plants and animals. The plants were forced to seek means to combat the insects, and the insects, in turn, to counteract these defenses. It was then that the drive for originality and adaptability began—a process that continues to this day. The emergence of a new defense mechanism in plants or a new adaptation in animals meant the appearance of a new biological species. Thus, the process of speciation in both kingdoms accelerated rapidly. Today, science agrees that the evolution of plants and herbivorous animals has proceeded in parallel.

The simplest method of combating herbivores is chemical. Plants began to produce toxins that are poisonous to animals. It is believed that the reasons for the success of angiosperms are their greater chemical flexibility and their enhanced ability to produce chemical defense substances. It is no coincidence that alkaloids, some of the most toxic compounds of plant origin, are found only in angiosperms. In addition to the spectacular geological and climatic changes that led to the massive die-off of ancient vegetation,

I. G. Ivanov, *The Invisible Language of Nature*,
https://doi.org/10.1007/978-3-662-73302-8_26

the main reason for the disappearance of ancient plant species is their poor protection against herbivorous animals and pathogenic microorganisms. Only those survived that began to develop protective mechanisms in time. Today's ferns, for example, are not as gigantic as their ancestors, but they produce substances with repellent effects. All modern plants release volatile substances into the atmosphere or non-volatile substances into the soil, with which they protect themselves from harmful microorganisms, fungi, and insects. It has been proposed to call these substances phytoncides (from Greek Φύτον, phyton—plant and Latin caedere—to kill). The term "phytoncides" was introduced in 1942 by the Russian microbiologist Boris Tokin, who was also the first to discover this previously unknown group of biologically active substances. He discovered phytoncides in 1930, when he observed that yeast fungi had difficulty developing or growing in the presence of onions. Tokin suspected that this was due to volatile substances released by the onion. Through a series of ingenious experiments, he proved that these substances not only killed yeast fungi but also many other pathogenic microorganisms. The volatile substances from garlic, parsley, mint, dill, anise, oak leaves, conifers, and others have such effects. In general, the release of phytoncides is a universal property of green plants. One hectare of deciduous forest releases about 2 kg of phytoncides per year, while the same area of coniferous trees (pine, fir, spruce) releases 5 kg. Juniper releases the most phytoncides, about 30 kg per year. The stronger emission of phytoncides from conifers is explained by a strong electric field around their needles, which promotes their dispersal in the air. The presence of such an electric field also explains the ability of conifers to ozonize the air. The high content of phytoncides in the air also determines the microbiological purity of mountain air. Its bacterial content is about 70 times lower than that of city air. The amount of bacteria in the air also decreases under the influence of volatile substances from domestic plants. For example, a pot of begonias or toadflax reduces the bacterial content of indoor air by 59%, while a chrysanthemum reduces it by 66%.

Under laboratory conditions, it has been shown that phytoncides have a broad spectrum of bactericidal effects. The phytoncides of onion and garlic, for example, kill almost all pathogenic bacteria as well as many protists (a group of unrelated microscopic organisms) in vitro. Therefore, since their discovery, great hopes have recently been placed on phytoncides for the treatment of infectious diseases. And indeed, how tempting is the idea of curing everything from angina to plague to cholera with garlic or onions. But disappointment was not long in coming. It turned out that phytoncides exert their bactericidal effect mainly outside the body, while they are ineffective inside the body. It is possible that they undergo chemical changes there. It is also possible that the pathogens themselves are better protected within the host organism. The healing effect of herbs and plant-based foods is associated with phytoncides. In the folk medicine of many countries, colds are still treated by inhaling the vapors of a decoction of essential plant oils (lavender, chamomile, basil, etc.), and in some Far Eastern countries, necklaces made of garlic cloves are worn as protection against infectious diseases. Onions are used to treat influenza, ear infections, contagious skin diseases, and so on. For eye and ear infections, folk medicine recommends chewing a clove of garlic and blowing the breath into the eye or ear of the patient. The phytoncides of garlic and onion have been further studied in modern medicine. From garlic, the antibiotic allicin was isolated (Fig. 1), which, even at a dilution of 1:250,000, kills streptococci, staphylococci, typhoid bacteria, paratyphoid viruses, tuberculosis bacilli, and so on.

Unfortunately, allicin is not stable and decomposes during storage. Numerous attempts have been made to use garlic under clinical conditions to treat acute catarrh of the upper respiratory tract, whooping cough, inflammations, ear and eye infections, dysentery, and purulent processes of the lungs. The results are not encouraging.

Phytoncides from many plants have insecticidal and repellent effects against insects, which is why they are used in agriculture to protect vegetable and fruit crops as well as to protect seeds from pests. For example, 200 g of garlic can protect 100 kg of wheat from fungus gnats (Sciaridae). In China, garlic is used in the storage of rice and flour. In Belgium, it is used to protect flaxseed from fleas. Aqueous extracts of garlic are also a proven remedy against spider mites.

The use of insecticidal plants is carried out in two ways: by planting them near crops or by spraying the crops with their extracts. For example, wormwood planted in orchards protects apples from the apple blossom weevil (*Caenorhinus aequatus*), and a decoction of wormwood kills leaf-eating caterpillars and acts as a repellent against blood-sucking insects and parasites in warm-blooded animals. The smell of hemp repels cockchafers, which is why it is recommended as an intercrop in orchards. In India, hemp leaves or whole hemp plants are used to repel fleas and bedbugs, and here, lavender is

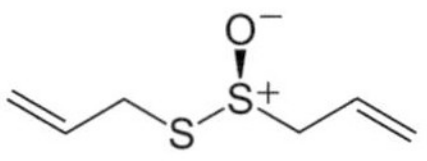

Fig. 1 Allicin

an old remedy against mealybugs and wax moths. Pyrethrum (Montenegrin or Dalmatian insect powder), obtained from chrysanthemums, also has insecticidal effects. Therefore, chrysanthemums are cultivated as industrial plants for the production of highly effective insecticidal preparations against insects and mites. They exert their toxic effect in amounts of 0.000017–0.000065% active ingredient, relative to the weight of the insect. They are absolutely harmless to warm-blooded animals and humans. The active ingredients are pyrethrin (Fig. 2) and cinerin (Fig. 3). These are contact insecticides, similar to DDT but with significantly stronger effects. They reach the nerve ganglia via the lymphatic system and cause paralysis and death of the insect. The commercial preparation "Pulvis Pyrethri" (pyrethrum powder) consists of the dried and powdered flowers of a Dalmatian chrysanthemum.

It is used to kill fleas and lice on hairy parts of the body. Pyrethrin and cinerin are representatives of the so-called polyene antibiotics, which are the main active ingredients of the phytoncides of composites. One of the simplest representatives of the acetylenic antibiotics is agropyrene (Fig. 4), which is found in couch grass roots.

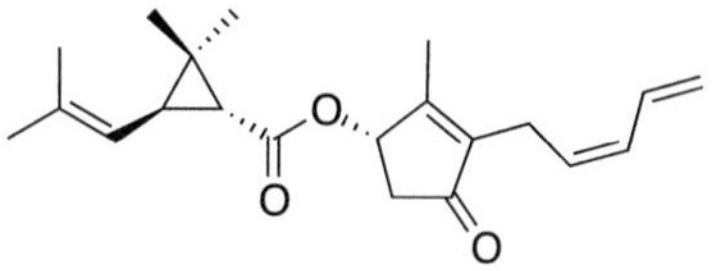

Fig. 2 Pyrethrin

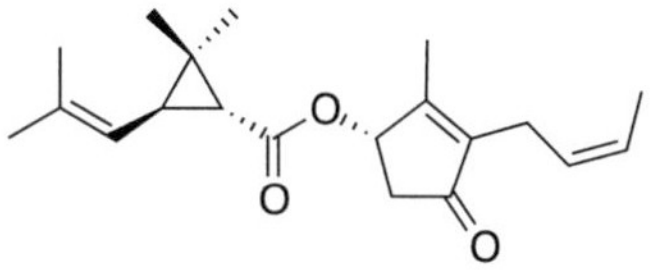

Fig. 3 Cinerin

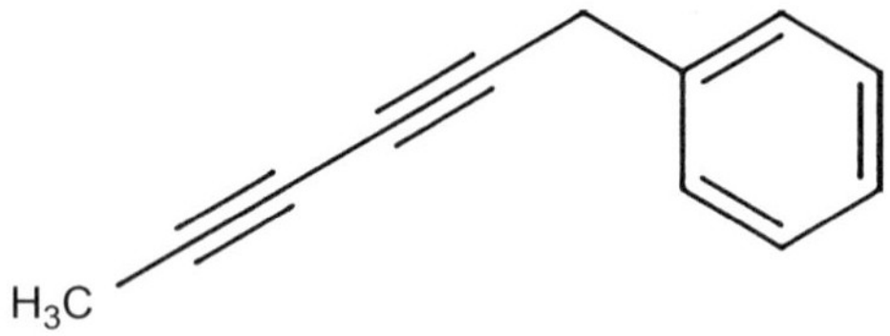

Fig. 4 Agropyrene

The volatile phytoncides of most plants have a positive effect on mammals and humans. The volatile emissions of mountain plants have a calming and strengthening effect on the nervous system. Prof. Holodin believes that some phytoncides, similar to vitamins, are essential for humans. In contrast to vitamins, however, they do not enter the body with food, but with the air we breathe. He calls them atmovitamins.

Phytoalexins

A description of chemical agents used to protect plants from their pests would be incomplete without mentioning phytoalexins. These are substances with bactericidal and fungicidal properties. They are produced when the plant's survival is threatened. These compounds are synthesized by the plant immediately after infection by microorganisms such as bacteria or fungi, in order to inhibit their spread, growth, or reproduction within the plant. The chemical nature of many of these substances has now been elucidated. It has been found that they belong to various chemical classes and groups. Most of them contain alcoholic or phenolic groups, which allow them to bind to carbohydrate residues in the form of glycosides. The glycoside form of phytoalexins is inactive; in healthy plant tissue, they are "dormant." If the tissue is damaged, specific enzymes (hydrolases) are activated, releasing the phytoalexin from its carbohydrate carrier. This kills the invading microbes or fungal hyphae. Thus, by attacking the plant, the pathogen itself triggers the trap that nature has "set" for it. Phytoalexins are the plant's second line of defense. The first are the phytoncides, which combat the pest from a distance. If they fail to stop it, the phytoalexins complete the task.

The study of the chemical structure of natural remedies is a highly attractive task for chemists. Through this research, many new compounds have been discovered, some with very interesting chemical structures. Phytoalexins are also of interest to biochemists because they exert their effects at various levels of cellular metabolism: they inhibit DNA, RNA, and protein biosynthesis, block respiration and oxidative phosphorylation, reduce the permeability of cell membranes, and so on.

I. G. Ivanov, *The Invisible Language of Nature*,
https://doi.org/10.1007/978-3-662-73302-8_27

In the search for effective means to combat pests, some plants have achieved astonishing originality. Dr. K. C. Williams of Harvard University discovered in the wood of the American balsam fir (*Abies balsamea*) the substance juvabione (Fig. 1), which is identical to the juvenile hormone of certain insects. The juvenile hormone regulates cell differentiation and development in insects during the larval stage. The transformation of the larva into an adult insect is controlled by another hormone, ecdysone (molting hormone) with a steroid structure. The effect of the juvenile hormone occurs at low concentrations. At higher concentrations, dramatic changes occur in the larval body, preventing it from developing into an adult. For this reason, juvenile hormones are also potent insecticides.

Despite the chemical defenses of plants, each of us has seen trees defoliated by caterpillars, young seedlings gnawed by rabbits, and beautiful pines, firs,

Fig. 1 American balsam fir (*Abies balsamea*) and the juvabione secreted by it, which is identical to the juvenile hormone of certain insects. (Photo: © Holcy/Getty Images/iStock)

and beeches riddled with holes by worms and beetles. This shows that herbivorous animals are by no means as helpless as the previous lines might suggest. Through evolution, almost every strategy is met with a counter-strategy. While plants have been equipped with perfect chemical defense mechanisms, animals have developed effective anti-chemical measures. Over the course of their evolution, animals have acquired the ability to produce enzymes that successfully break down and eliminate the toxins intended for them.

As we have already seen, natural chemical substances are of diverse chemical nature. This means that their elimination cannot follow a single scheme. Each toxic substance requires a specific neutralization pathway, but maintaining a large metabolic network is energetically disadvantageous for the individual. For this reason, herbivorous animals are not generalists, but adapt to a single or a few plant species.

For example, the caterpillars of some butterfly species in the family Papilionidae feed on apple leaves, others on magnolia leaves, still others on citrus plants and parsley. The monarch butterfly *Danaus plexippus* (Fig. 2), on the other hand, prefers milkweed (*Asclepias*).

Many insects recognize their favorite food precisely by the scent of the substances it contains. For example, the onion fly *Delia antiqua* locates its food by the smell of propyl mercaptan and propyl disulfide. Butterflies of the Pieridae family (whites), which feed on cruciferous plants, recognize them by the isothiocyanates, which strongly irritate the mucous membranes of mammalian eyes. The caterpillars of the butterfly *Papilio ajax*, which feed on the leaves of umbellifers, are attracted by the secreted carvone and methyl chavicol (Fig. 3). If filter paper is soaked with these substances, they immediately begin to nibble on it. The cucurbitacins of the gourd family (Cucurbitaceae), in turn, attract the corn rootworm (*Diabrotica*

Fig. 2 *Danaus plexippus.* (© olikli/Getty Images/iStock)

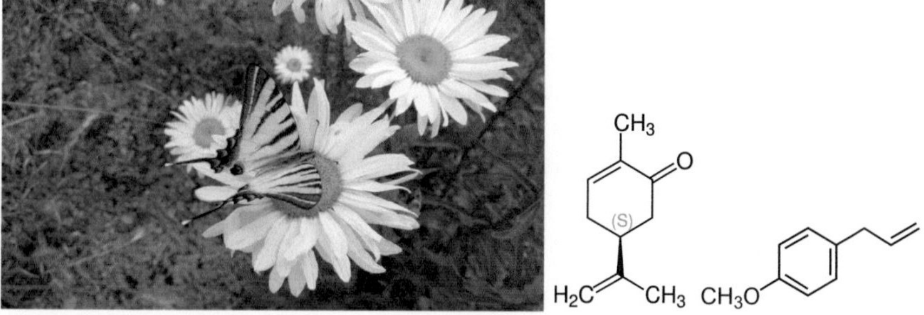

Fig. 3 The butterfly *Papilio ajax* and its attractants carvone and methyl chavicol. (Photo: © Ellita/Getty Images/iStock)

undecimpunctata) and repel bees as well as wasps. Some plants are vital for insects because they contain substances unique to them. For example, bees collect pollen from clover not only for the pollen itself, but also because they obtain octadecatrienoic acid from it. This is the raw material for the synthesis of the queen substance.

In our previous discussion of the relationships between plants and other living beings, we have portrayed animals only as enemies of plants. However, we know that without insects, most flowering plants would disappear from the earth. They are involved in a crucial phase of their life cycle—pollination. To attract beneficial insects, plants use chemical agents that are no less effective and ingenious than those used for defense or destruction. During flowering, they secrete chemical substances that exert a strong attraction on insects. In nature, these sometimes resemble pheromones. Some mimic aggregation pheromones, others trail-marking pheromones. There are even true sex pheromones known.

For example, orchid flowers of the genus *Ophrys* contain alpha-copaene and alpha-cadinene (Fig. 4), which act as female sex pheromones for *Gorytes mystaceus* and *Gorytes campestris* (species of digger wasps). Attracted by the scent, the males land on the orchid flowers and become aroused. They attempt to mate and, through this deception, carry out pollination.

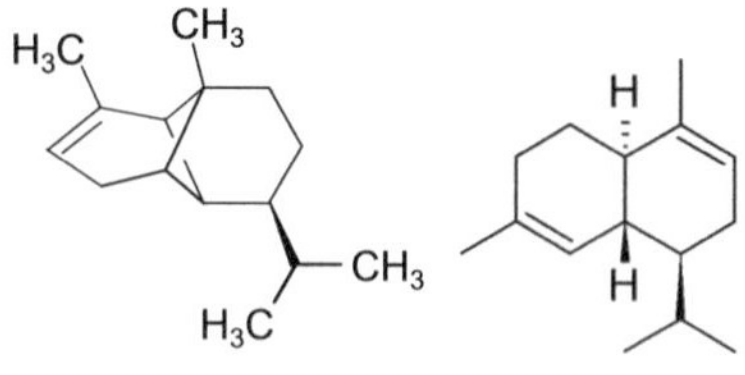

Fig. 4 Alpha-copaene and alpha-cadinene

Predatory Fungi and Nematodes

In addition to pollination, insects also serve as food for plants in some cases. More than 450 species of lower and higher predatory plants are known, whose relationship to their prey is based on chemical mechanisms.

The predatory fungi belong to the order of Hyphomycetes (thread fungi). Some of them are also assigned to other taxonomic groups (Zygomycetes). They live mainly in water and feed on microscopic animals such as nematodes (roundworms), protists, and insects. However, their favorite food is nematodes. These are worms measuring 0.1–1 mm in length, distant relatives of the ascarid worms. They live in natural water sources and feed mainly on organic debris that accumulates at the bottom of bodies of water. Many nematodes are pathogenic to both animals and plants.

The mycelia (filamentous cells of a fungus) of predatory fungi develop on plant debris but feed mainly on animals. For them, the body of the victim is, as for predators, merely food and not a habitat as it is for parasites. They generally feed on animals that are significantly larger than themselves. While the length of the roundworms, for example, reaches 1 mm, the thickness of the fungal hyphae is only 5–8 μm (0.005–0.008 mm). Capturing such large, dangerous, and strong prey is only possible thanks to the subtle and sophisticated hunting devices developed over the course of evolution. The most important hunting tool is the trap. These are constructed differently in predatory fungi. The simplest is the adhesive trap. This consists of undifferentiated, branched hyphae coated with a sticky substance. This is capable of immobilizing and fixing a roundworm (Fig. 1). The sticky net often reaches a considerable size, increasing the likelihood that a victim will fall into it.

I. G. Ivanov, *The Invisible Language of Nature*,
https://doi.org/10.1007/978-3-662-73302-8_28

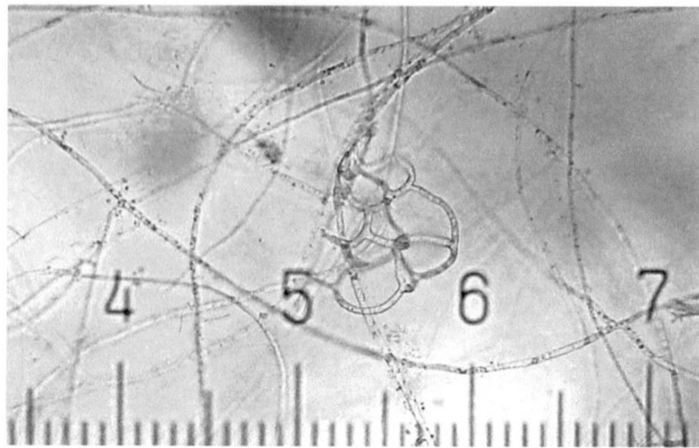

Fig. 1 Adhesive traps of a predatory fungus (Bob Blaylock, CC BY-SA 3.0, Wikimedia Commons)

When the worm touches the net, it becomes stuck. As it tries to free itself, it becomes more and more entangled, like a fly caught in a spider's web. Its movements become increasingly frantic, and eventually the worm dies.

It is likely that toxins contained in the sticky fluid (called nematoxins) also contribute to the death of the victim. After capturing the prey, a hypha proliferates and penetrates the body of the roundworm. Inside, trophic (nutritional) hyphae grow and fill the interior of the worm. They extract its vital fluids for about a day, and in the end, only the outer shell with the trophic hyphae remains. If the victim is strong enough to break free, it is still doomed, as it takes adhering hyphae with it, which will lead to the same result.

The traps of other predatory fungi operate purely mechanically. They consist of three cells arranged in a circle that close around the mycelium. In the simplest case, they act passively. When a nematode tries to enter the circle, it is mechanically trapped inside. The best-studied traps are of the "constricting ring" type. Externally, they resemble passive traps, but differ in that their inner wall is sensitive to touch. Upon contact, the cells of the ring swell within 0.1 s and assume an almost spherical shape, so that the interior of the ring is filled. The swelling of the cells is irreversible. The roundworm is tightly gripped in the trap. The force of contraction is sufficient to kill the prey purely mechanically. Further consumption proceeds in the same way as with adhesive traps.

By studying the behavior of predatory fungi and roundworms, science has concluded that their interactions are based on complex chemical communication. The mere entry of nematodes into traps is not a random event.

In model experiments with artificial traps whose parameters match those of predatory fungi, significantly fewer nematodes are caught than in natural traps. It is assumed that the preference for natural traps is due to attractants secreted by the hyphae. The nematodes, in turn, also influence the development of traps. This assertion is based on the fact that predatory fungi can be easily cultured in vitro, but do not form traps in the absence of nematodes. However, if nematodes are added to the nutrient medium, trap formation occurs within just 24 h. Traps also form when water in which nematodes have previously lived is added. This shows that the stimulus does not come from the worms themselves, but from chemical substances they secrete. The fact that nematode-free water retains its stimulating effect even after sterilization at high temperatures indicates that the biologically active substances are heat-resistant compounds. Most likely, these are amino acids or short peptides, called nemin.

Carnivorous Plants

These plants are also green and no less delicate and beautiful than others, but unlike them, they love meat. In the plant world, they are true "wolves in sheep's clothing." The first description of an insect-eating plant dates back to 1769, and in 1875, "Insectivorous Plants" by Charles Darwin was published. At the time, it was fiercely criticized by botanists. The director of the Pittsburgh Botanical Garden wrote: "This is a theory at which any reasonable botanist would laugh. If the English believe it, it is only out of respect for Mr. Darwin."

Most carnivorous plants live in swampy areas where there is a lack of nitrogen. They have a moderately developed root system, and some are even completely rootless. The nitrogen deficiency has forced them to include meat in their diet. In this context, it is not uninteresting to mention that cannibalism, which was practiced in the past in certain regions of New Guinea and the Pacific Islands, is also, according to some hypotheses, due to nitrogen deficiency.

Carnivorous plants, like other green plants, carry out photosynthesis, i.e., their main carbon source is carbon dioxide. Due to the lack of inorganic nitrogen, they have developed the ability to utilize organic (animal) nitrogen. For this purpose, carnivorous plants have developed ingenious hunting devices, in which chemistry also plays a role. The following will consider two typical representatives of predatory plants, sundew and the vines from the pitcher plant family (Nepenthaceae). *Drosera rotundifolia* (Fig. 1) grows in peat bogs as a small, delicate plant with rosette-like leaves. Above the rosette, small white flowers rise on long, thin stems, covered with tiny, transparent,

I. G. Ivanov, *The Invisible Language of Nature*,
https://doi.org/10.1007/978-3-662-73302-8_29

Fig. 1 Round-leaved sundew *Drosera rotundifolia.* (© Juan Francisco Moreno Gamez/Getty Images/iStock)

honey-scented droplets of liquid. They resemble morning dew. In reality, it is a sticky liquid that is part of the dew drop trap. When an insect, attracted by the pleasant scent and appetizing appearance of the droplets, lands on the rosette, it gets stuck. Numerous glands are activated and begin to secrete copious amounts of mucus, which envelops the insect from all sides. It soon suffocates, and digestion can begin. The secretion is a true digestive juice. It contains proteolytic enzymes, similar to the stomach enzymes pepsin and trypsin, which break down proteins into amino acids. Once the insect is digested, the petals open and the wind cleans the trap of the chitinous remains of the victim. Then new dew drops appear and the trap is ready for a new digestive cycle.

The dew drop reacts not only to insects but also to pieces of meat. Darwin wrote that his sundew had a preference for rump steak. In experiments to determine the sensitivity of the sundew to stimuli, it was shown that it even reacts to 0.000822 mg of meat. The trap is not only sensitive to touch but is also triggered by the smell of meat. It does not respond to plant stimuli. This means that the sundew is not an omnivore. Moreover, it has a preference for certain insects. The dew leaf (*Drosophyllum lusitanicum*) for example, feeds only on flies, which is why it is sometimes kept as a pet flycatcher.

The hunting equipment of the predatory vines from the Nepenthaceae family is based on a completely different principle (Fig. 2). They inhabit the tropical forests of the islands of Borneo, New Guinea, and Madagascar. Their trap is a modified, cone-shaped leaf with a diameter of up to 30 cm and a height of up to 50 cm. The rim of the cone is covered with numerous

Fig. 2 Cone of a vine from the *Nepenthaceae* family. (© trutenka/Getty Images/iStock)

nectar glands that secrete a highly aromatic nectar attractive to insects. The glands are arranged so that the aroma intensifies from the periphery toward the interior of the cone. This entices insects to enter, until they eventually fall into the interior of the trap. Since the inner wall is covered with a smooth, waxy layer and the cone itself is filled up to half its volume with liquid, it is practically impossible to escape. The liquid is a digestive juice, similar to that of the sundew. It is secreted by special glands, whose number is 6000/cm^2. In addition to proteolytic enzymes, the digestive juice also contains formic acid, which acts as a preservative and protects against decay. To prevent the juice from being diluted by rainwater, the cone is closed with a lid. Although ants and cockroaches are the main victims of these vines, sometimes small rodents and birds, which are hunting insects, also fall into the trap. When the cones become unusable, they dry out and new ones form in their place.

How did predatory plants originate? It is assumed that they originated in swamps, where there have always been carcasses of insects, crustaceans, and other small animals. At first, dead animals served as food for saprophytic (feeding only on dead organic matter) plants, but gradually some plants also learned to catch live prey themselves. The fact that plants can, under certain conditions, feed in a manner similar to animals speaks to the close relationship between the plant and animal kingdoms.

The Secret of Oak Galls

We have become so accustomed to some of nature's creations that it seems pointless to inquire into their true nature. This is also the case with oak galls (also called oak apple, oak gall, leaf gall, or oak apple, Fig. 1). Who has not seen them at some point, and who has not played with them as a child? Yet how many of us actually know what oak galls are? They are an excellent example of the complex interrelationships between plants and animals.

In our region, oak galls are found mainly in oak forests. Although they resemble nuts or acorns, hardly anyone would think they are fruits. If you look closely at a ripe oak gall, you will notice a small hole on its surface, similar to that on a worm-eaten fruit. When opened, you will find traces of a worm inside the gall. But you will not find a worm. Any attempt to find a ripe gall without a hole or with a living worm will be unsuccessful. Only young oak galls, still soft and juicy like a strawberry, have no holes. If you cut one open, you will find a worm inside. This raises the question: How did the worm get inside without leaving a hole? For many years, this was a mystery to science. At first, it was thought that the worm fertilized itself inside the gall. Later, it was assumed that its eggs were sucked up from the soil through the oak's roots. The precise answer to this question was only provided much later, after patient and thorough study of the worm's life cycle. It turned out that there is a close relationship between the worm and the oak gall.

The "worm" is the larva of a small wasp, *Cynips gallae,* known as the gall wasp (Fig. 2). It belongs to the order Hymenoptera and is a member of the family Cynipidae (gall wasps). In spring, the wasp lays its eggs on buds or

I. G. Ivanov, *The Invisible Language of Nature*,
https://doi.org/10.1007/978-3-662-73302-8_30

Fig. 1 Oak galls. (© esemelwe/Getty Images/istock)

Fig. 2 *Cynips gallae.* (© FrankRamspott/Getty images/iStock)

under the leaves of the oak. Sometimes it also lays eggs under the bark of young twigs, which it pierces with its sharp ovipositor. The site of egg deposition is always near large, sap-conducting vessels. The egg develops into a larva ("worm"), which secretes chemical substances that strongly stimulate cell division. At the point of contact between the larva and the oak, a local proliferation of plant tissue begins. The small larva remains at the center of this growth. The swelling quickly grows to the size of a ripe fruit, initially soft and juicy, later hardening. The mass of the fruit is sufficient to nourish the voracious larva until it becomes an adult insect. Upon reaching maturity, the young wasp breaks through the fruit's shell and emerges. This is how the hole in the oak gall is formed. In other words: the fruit is a chemically induced benign tumor. By inducing tumor-like growths, the chemicals secreted from the eggs provide comfort, nutrition, and good health until fully developed winged insects emerge, without causing any significant harm to the host tree.

Sometimes the wasp also lays its eggs on other tree species. When eggs are laid on young trees, unsightly swellings (tumors) form on the branches and trunk, which remain for the tree's lifetime. In addition to insects of the Cynipidae family, insects from the families Tenthredinidae (sawflies) and Cecidomyiidae (gall midges) also produce similar growths. Their shapes vary; they can be spherical, elliptical, smooth, rough, with or without rosettes, multilayered, double-layered, etc. It appears that the shape is specific to each species, which in turn illustrates species specificity and the different effects of the chemicals secreted by the insects. These are not just growth inducers, but substances that contain specific information about the architecture of the induced tumor. Some believe that the wasp introduces an oncogenic virus into the plant along with the egg. However, this has not been proven. It is also not fully understood whether the wasp or the larva secretes the biologically active substances. The view that the wasp itself triggers the growth is supported by two facts: First, the growths do not differ from those induced by a living larva if the larva is destroyed by a parasite in the early stages of development. Second, in some species (e.g., *Pontania*), an egg is not necessarily required for the formation of a swelling.

The molecular mechanisms of the processes leading to the formation of oak galls are not yet fully understood. Little is known about the chemical nature of the substances released by the insects. Their quantity is too small for detailed chemical analysis, and the larvae or gall wasps cannot be cultured in the laboratory. Unraveling the mystery of this unique natural phenomenon is one of the many challenges facing future naturalists.

Chemical Weapons of Animals

In the chemical laboratories of animals, substances are produced not only for informational purposes but also for chemical defense. In contrast to pheromones and musk, which benefit both the producer and the recipient, toxins are advantageous only to the producer. Since they are intended to affect organisms of another species, they are officially classified as allomones.

The term toxin originally referred to the poisons used by ancient peoples to make poisoned arrows. It was later defined much more broadly. Prefixes such as "myco-" (fungus), "phyto-" (plant), "neuro-" (nerve), etc., were introduced to specify the various types of origin or effect. In the following, we use the term "toxin" as a synonym for poisonous substances.

There are about 50,000–60,000 poisonous animal species on Earth, with representatives in almost all taxonomic groups. Their number is greatest among arthropods, which account for 80% of all poisonous animals. The smallest number of poisonous animals is found among mammals. In fact, only a single truly poisonous mammal is known to date, and that is the platypus.

The chemical weapons of animals differ significantly in their equipment and mode of action. Some venomous animals possess specialized venom glands, usually connected to organs for wounding and injecting the venom into the body of the victim or attacker. These are the so-called actively venomous animals, which include many species dangerous to humans. Others have venom glands but lack an apparatus for inflicting injury (passively venomous animals). In others, the venom glands are combined with mechanisms that allow the venom to be sprayed toward the opponent. There are

I. G. Ivanov, *The Invisible Language of Nature*,
https://doi.org/10.1007/978-3-662-73302-8_31

also species known that have no specialized venom glands at all. In terms of their biological effect, animal venoms can be simple defensive substances, cytotoxins, or neurotoxins, and their chemical composition ranges from simple organic compounds to complex cocktails of enzymes, peptides, and proteins.

The simplest form of chemical defense is to make oneself unpalatable, i.e., poisonous or unpleasant-tasting. Venom glands are not even required for this. It is sufficient if the organism saturates its body with toxins from its food, to which it is itself resistant. This is the case with the monarch butterfly (*Danaus plexippus,* Fig. 1, left), whose caterpillars feed on milkweed plants of the genus *Asclepias*. These are rich in cardiac glycosides (cardiotoxins). Cardiac glycosides taste bitter and are highly toxic to vertebrates, but not to insects. They are not broken down but accumulate in large quantities in both the caterpillars and the butterflies. When a bird pecks at a monarch butterfly or its caterpillar for the first time, it initially eats it but then quickly spits it out. This creates a lasting conditioned reflex, which is also reinforced by the butterfly's bright coloration. The bird remembers both the butterfly and its caterpillars for life and never picks them up again. To prove that the toxins in *Danaus plexippus* come from milkweed plants, the American entomologist and ecologist Lincoln Brower raised monarch butterflies on cabbage and demonstrated that they were then non-toxic to birds. The birds' aversion to the monarch butterfly has led some non-toxic butterflies, such as the swallowtail *Papilio dardanus* (Fig. 1, right), to develop coloration similar to that of the monarch. Similar mimicry has also been observed in other butterflies.

Fig. 1 Monarch butterfly (*Danaus plexippus*, left, © olikli/Getty Images/iStock) and swallowtail. (*Papilio dardanus,* © johnandersonphoto/Getty Images/iStock)

A similar situation is found with the aphid *Aphis nerii*, which accumulates cardiotoxins derived from oleander in its body to protect itself from predators. The honeybee also benefits from plant toxins. It produces propolis, which it uses to protect the hive from harmful microorganisms, mold, and fungi. Laboratory experiments have shown that propolis kills more than 100 species of pathogenic microorganisms. Chemical analyses indicate that the bactericidal effect of propolis is due to its phenol and terpene compounds. These include benzoic acid, p-hydroxy- and p-methoxybenzoic acid, p-coumaric acid, isovanillin, the terpenoids acetoxybetulinol and bisabolol, as well as the flavonoids kaempferol, naringenin, etc. The raw materials for the production of propolis are obtained from tree buds, which are particularly rich in such compounds. Therefore, propolis is produced before the trees bloom and not during the collection of nectar and pollen. Bulgarian scientists led by Vasya Bankova demonstrated that Bulgarian propolis originates from the black poplar.

Another example of passive chemical defense is toxic hemolymph (the blood fluid of insects), regardless of the food the insects consume. This includes, for example, beetles of the genus *Mylabris* (Fig. 2). As with most poisonous organisms, their coloration is bright and vivid. They have a black head and thorax, as well as black or orange wings with black spots. All of this serves as a warning of danger and is essential for the beetle's protection. In dangerous situations, the *Mylabris* beetle secretes droplets of toxic hemolymph through a small opening between the knees and thighs. If an inexperienced bird pecks at such a beetle, it spends a long time cleaning its

Fig. 2 Oil beetle of the genus *Mylabris*. (© Henk Wallays/Getty Images/iStock)

beak on the ground. If, on the other hand, a horse, a camel, or another herbivorous animal swallows such a beetle, it can develop enteritis, which may be fatal. For this reason, livestock farmers are particularly careful to ensure that their animals do not graze on pastures infested with *Mylabris* beetles. They are also dangerous to humans, as they can cause skin inflammation and blistering similar to burns.

The secretion of hemolymph is also observed in praying mantises. When threatened, they secrete hemolymph from their legs, whose unpleasant odor is perceptible to humans. This and the bad taste of the hemolymph are due to quinones, which are repellent to insects and toxic to higher animals. Some insects, such as the American *Cimbex* wasp (birch sawfly) and certain species of grasshoppers, are able to spray toxic hemolymph in a thin jet from special openings above the tracheae.

Another widespread form of harmless chemical defense is the release of foul-smelling substances. These are produced by glands located at various points on the animal's body. The foul-smelling secretions of the common bedbug, for example, are found on the first three segments of the body, or, as in the case of the stink bug, on the abdomen and back. The secretions from these glands are not only repellent to most insects but also act as powerful contact insecticides. After contact with these glands, insects become paralyzed and may die.

An unpleasant-smelling liquid is also secreted by the ant *Tapinoma erraticum*. This allows them to move freely among insects much larger than themselves. Its odor can be detected several meters away. In some odor-sensitive people, it causes nausea. All in all, ants are among the most chemically sophisticated organisms. In our study of pheromones, we found that they secrete numerous informative and chemically diverse compounds. But the chemical defense substances they produce are no less diverse. For example, the red wood ant (*Formica rufa*) secretes a substance containing 20–70% formic acid. The secretion of *Myrmicaria natalensis* (Natal droptail ant) consists of 35% acetic acid, 31% isovaleric acid, 22% propionic acid, and 12% water. The Argentine ant *Linepithema humile* produces a secretion with a strong insecticidal effect. The main component is iridomyrmecin (Fig. 3). Another insecticide produced by the ant *Dendrolasius fuliginosus* is dendrolysin. The ants *Tapinoma nigerrimum, Iridomyrmex rufoniger, Dolichoderus scabridus*, and others secrete a substance containing the dialdehydes iridodial and anisomorphal, which are strong lacrimators (tear-inducing substances) for humans and mammals. The venom of the bulldog ant (*Myrmecia gulosa*) contains histamine, hyaluronidase, and hemolysin.

Fig. 3 Argentine ant (*Linepithema humile*, former name *Iridomyrmex humilis*) and its toxin iridomyrmecin. (Photo: © Heather Broccard-Bell/Getty Images/iStock)

As for the venom apparatus, there are stinging and non-stinging ants. Stinging ants deliver their venom in the same way as bees and wasps. Non-stinging ants first injure the tissue with their strong mandibles, then turn their bodies and inject the secretion from their anal glands into the wound. In addition to the chemical compounds used for communication and defense, ants also produce numerous other substances with various effects. For example, the leafcutter ant *Atta sexdens*, which cultivates a specific fungus species in its nest, secretes phenylacetic and indoleacetic acid as well as the fungicide myrmicacin. Phenylacetic acid protects the nest from the development of pathogenic bacteria, indoleacetic acid stimulates the growth of the fungi, and myrmicacin is a fungicide that inhibits the development of fungi other than those cultivated. Fungicides are also secreted by the harvester ant *Messor barbarus*. They collect and store seeds from cereal plants, which they spray with secretions from their anal glands to protect them from mold.

Some passively poisonous animals, i.e., those without organs for inflicting injury, literally carry their poison on their backs. These include frogs, salamanders, and newts. On the surface of their bodies are two types of glands: mucous and serous. The former keep the skin moist, and the latter produce toxic substances. The venom of the fire salamander (*Salamandra salamandra*, Fig. 4) is one of the first animal toxins to be discovered. As early as 1866, it was established that the secretion from the salamander's serous glands has the empirical formula $C_{68}H_{60}N_{20}O_{10}$.

This substance is characterized by a strong neuroparalytic effect and is called salamandrin. If a salamander is placed in an aquarium with fish, after a few minutes the coordination of the fishes' movements collapses, followed

Fig. 4 Salamander (*Salamandra salamandra*). (© Tree4Two/Getty Images/iStock)

by paralysis and death. If a salamander is bitten by a viper, the viper dies after about 4 hours. However, if a viper eats a salamander, it dies after 4 minutes. Salamander venom is also dangerous to warm-blooded animals.

The venom of some frogs is even more potent. Indigenous peoples of South America have long used certain species of poison dart frogs to make their poison arrows. To always have fresh venom at hand, they even carried live poison frogs in special containers. They would rub the tips of their arrows with the frogs' skin before use. The toxin of one of the most poisonous poison dart frogs contains batrachotoxin (Fig. 5). Like that of the salamander, it is neuroparalytic.

The toxic secretions of amphibians also have bactericidal effects. This is not surprising. The skin of amphibians is always moist and provides a good environment for the growth of microorganisms. Without protection against infections, the animals would quickly perish. It has been experimentally demonstrated that amphibians die of skin inflammation if their toxin is neutralized.

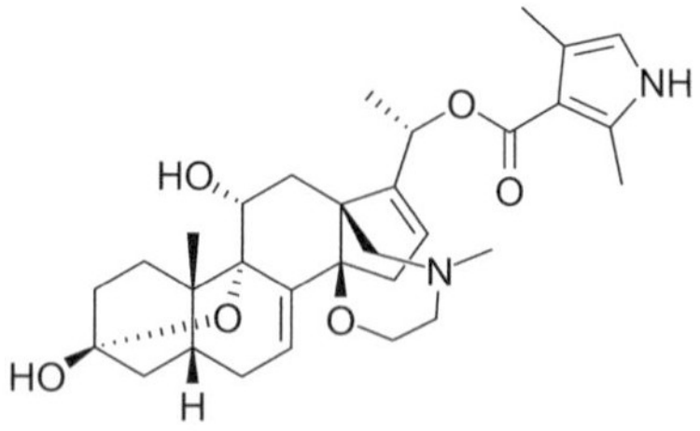

Fig. 5 Batrachotoxin secreted from the skin of some tropical poison dart frogs

The Deadly Weapon of Snakes

No animal has played as significant a role in religion and superstition as the snake. Since ancient times, it has inspired fear and respect in humans. Helplessly exposed to the snake's venom, people regarded it either as a divine being or as the embodiment of the devil. In ancient Greece, the snake was a symbol of wisdom, and in India, the cobra is still revered and respected today. A Hindu feels honored if a cobra enters his house. He does not kill it, but waits for it to leave voluntarily. According to Hindu legends, the spots painted by Buddha on the neck of the spectacled cobra are eyes, meant to frighten away kites (birds of prey) and thus protect their offspring (Fig. 1). In addition to fear, snakes have always aroused the curiosity of naturalists.

There are two reasons for the interest in snakes: first, to obtain information about their venom with the aim of finding a suitable antidote; second, to explore the possibility of using the antidote as medicine. Pliny the Elder and, after him, other thinkers of antiquity proposed a long list of remedies made from various snake organs. Well known is theriac (an antidote to animal poisons, especially snake venom), a complex mixture of herbs and snake venom. It was developed in the first century AD, remained in use until the end of the 18th century, and was listed in the official English pharmacopoeia until 1746. It is no coincidence that the symbol of medicine and pharmacy to this day is the snake.

I. G. Ivanov, *The Invisible Language of Nature*,
https://doi.org/10.1007/978-3-662-73302-8_32

Fig. 1 Spectacled cobra (*Naja naja*). (© shylendrahoode/Getty Images/iStock)

Which Snakes Are Venomous?

Of the approximately 4000 known species of snakes, about 700 are venomous. However, the figures in the literature are inconsistent. Most venomous snakes are found in South America, with the fewest in Europe. Venomous snakes are divided into elapids (Elapidae) and vipers (Viperidae). Elapids include, for example, sea snakes (Hydrophiinae) and cobras (*Naja*), while vipers include rattlesnakes (*Crotalus*) as well as, in our region, adders and common European vipers (*Viperinae*). Venomous snakes are not very large; their length rarely exceeds 100–150 cm. The largest venomous snake is the king cobra, which is native to Southeast Asia and reaches a length of 3.3–3.8 m (Fig. 1). A record length of 4.57 m has been reported.

Without exception, all venomous snakes possess venom glands for producing venom as well as fangs for wounding and injecting the venom into the victim's body. Their fangs are hollow and located in the upper jaw. Venom glands are modified salivary glands and are symmetrically positioned behind the eyes. They are enclosed by a capsule of fibrous tissue and have the shape and size of an almond kernel. Muscles and tendons are attached to the glands, serving to squeeze out the venom when the fangs penetrate the victim (Fig. 2).

The fangs are sharp, saber-shaped, and lie concealed behind the fold of the gum tissue. When the mouth is closed, they are horizontal, with the pointed end facing inward. When the mouth opens, they snap forward and retract slightly. Inside the fangs is a channel that transports the venom from the glands to the wounded tissue. In addition to the two main fangs, smaller reserve fangs can be seen in the upper jaw, which replace the main fangs if lost.

I. G. Ivanov, *The Invisible Language of Nature*,
https://doi.org/10.1007/978-3-662-73302-8_33

Fig. 1 King cobra (*Ophiophagus hannah*). (© Cavan Images/Getty images/iStock)

Fig. 2 The venom apparatus of the common European viper *Vipera berus*. (© Nastasic/Getty Images/iStock)

In some snake species, the venom apparatus is arranged so that venom can be ejected up to 2 m toward the victim. These are the "spitting" snakes. Even the first Europeans who visited Asia and Africa told terrifying stories of snakes that could blind a person from several meters away. At the time, no one believed these stories, although they were emphatically confirmed by the local population. Later, naturalists also described these fearsome animals.

Fig. 3 Spitting cobra. (© Michal/Stock.adobe.com)

They turned out to be spitting cobras. Several species of spitting cobras are known. They are widespread in Africa and Asia. Among the best known are the red spitting cobra (*Naja pallida*), the ring-necked spitting cobra (*Hemachatus haemachatus*), and the Javan spitting cobra (*Naja sputatrix*). Many spitting cobras can grow very large, reaching body lengths of about 2 m. Like other venomous snakes, they kill their prey by biting. The venom spray serves as protection against large predators. The startled spitting cobra assumes a characteristic threat posture. It holds its head horizontally, raises the front part of its body vertically, and spreads its hood. Then it takes aim and, with enviable accuracy, shoots a thin jet of venomous fluid into the eyes of the aggressor (Fig. 3). This is followed by rapid inflammation of the conjunctiva, clouding of the cornea, and thus loss of vision. The pain is so intense that the victim is literally paralyzed by it. What is the mechanism behind the venom spray? The venom apparatus of spitting cobras is constructed somewhat differently from that of other venomous snakes. The venom glands are located in the middle of the head and are connected to strong muscles, whose contraction increases the pressure in the gland to 1.5 bar. This creates the conditions for the venom to be ejected as a jet from a distance of 1.5–2 m. Two jets are fired, which merge into one at a distance of 0.5 m from the head (Fig. 3). An agitated snake can deliver up to 28 consecutive shots. The time for a single shot, from opening to closing the mouth, is only a quarter of a second, with the venom being expelled 0.07

s after the mouth opens. Due to the lightning-fast speed, the attacker does not even have a chance to blink, and the venom goes directly into the eye. To improve aiming accuracy, the snake's trachea closes during ejection. This prevents the jet from being dispersed by exhaled air.

How Potent Is Snake Venom and How Dangerous Is It for Humans?

Snake venom is only lethal when injected into subcutaneous tissue or a blood vessel. If applied to the skin or external mucous membranes, it causes inflammation but does not result in poisoning or death. Mortality is highest with intravenous injection, lower with intramuscular injection, and lowest with subcutaneous injection. When ingested orally, it is virtually ineffective. The lethal dose of venom varies among different snake species. Some of the most venomous snakes are cobras. One gram of cobra venom can kill 25 dogs, 60 horses, 165 humans, or 300,000 pigeons. Sea snakes are seven to eight times more venomous than these. The vipers and adders native to our region are among the less venomous snakes, although they should not be underestimated.

The consequences of a venomous snake bite depend both on the potency of the venom and the amount injected. These two parameters, in turn, depend on several other factors, such as the age and sex of the snake, the season, the life cycle (before or after shedding), physiological state (hungry or satiated, calm or agitated), and so on. It has been experimentally demonstrated that the toxicity of the venom can increase up to tenfold after prolonged starvation and after shedding. Different animals have varying sensitivities to snake venom. Horses are among the most sensitive, while rabbits and dogs are significantly more resistant. While 1 g of cobra venom kills 30,000 kg of horses, the same amount can kill 250 kg of dogs, 2000 kg of rabbits, and 1430 kg of rats.

Some animals have innate immunity to snake venom. This allows them to include venomous snakes in their diet when the opportunity arises. These

I. G. Ivanov, *The Invisible Language of Nature*,
https://doi.org/10.1007/978-3-662-73302-8_34

include, for example, the hedgehog, mongoose (family Herpestidae), fox, pig, and skunk. The way they kill and eat snakes varies. While hedgehogs and skunks bite through the snake's neck and eat only the internal organs, the fox grabs the snake's head and slams it against the ground until it stops moving. Then it eats the head first, followed by the rest. It has been observed that a fox can devour three large snakes in an hour. Dogs also attack snakes, but since they are not immune to the venom, their encounters with snakes can be fatal.

Statistics on snakebite victims are quite contradictory. One reason for this is that in poorer countries, where most venomous snakes are found, statistics are poorly maintained. There are reports that up to 150,000 people die from snakebites worldwide each year. However, some epidemiologists believe these numbers are underestimated. In many Third World countries, unexplained deaths are often attributed to snakebites. A large proportion of snakebite victims are not included in the statistics because there are often too few hospitals to collect such data. The number of bites is, of course, much higher, but by no means do all end in death. In many cases of venomous snakebites, death does not occur even without medical intervention. The reason for this is a fortunate combination of circumstances—good health of the bitten person, poor physical condition of the snake, superficial penetration of the venom, and so on.

According to their mode of action, snake venoms are divided into two groups: vasotoxic and neurotoxic venoms. The first group includes the venoms of Viperidae (vipers, adders) and rattlesnakes, the second those of sea snakes and cobras. The venom of the former acts locally and produces edema within minutes, which quickly spreads to cover the entire limb. Due to irreversible damage, necrosis develops in the soft tissue. The permeability of blood and lymph vessels increases sharply, making them permeable to blood plasma and erythrocytes. Post-traumatically, there is severe bleeding in the affected limb as well as in the gastrointestinal tract, lungs, tissue around the heart, and other organs. As a result of blood loss, which can reach up to 50%, acute anemia develops. Superficial bleeding also contributes to this. Incidentally, the venom of vipers and rattlesnakes causes a two-stage effect on the blood coagulation system. A few minutes after injection of the venom, there is accelerated hemocoagulation (clotting). This leads to thrombus formation. After 1–3 hours, fibrinolysis occurs, and the blood loses its ability to clot. The bite also causes a loss of elasticity in the blood vessels, accompanied by obstruction and blood congestion. This affects not only the limbs but also the blood vessels of internal organs, resulting in edema and dystrophic changes. Watery blisters form on the limbs, which burst. In their place, ulcers develop that are slow to heal. Overall, these are vasotoxic reactions.

The clinical picture is quite different with bites from some rattlesnakes and sea snakes. The most characteristic symptom here is rapid and progressive paralysis. Initially, the bite site is painful, but there is little or no edema. Soon, coordination and movement are impaired, the bitten limb becomes paralyzed, and the victim becomes increasingly immobile. The paralysis quickly spreads to other muscles of the body, including the respiratory muscles. At the same time, the glottis is paralyzed, the soft palate becomes flaccid, and breathing becomes difficult. The tongue becomes numb, and salivation begins. The saliva cannot be swallowed and flows into the airways. Breathing ceases while consciousness is fully preserved; the heart continues to beat, and the pupils may still react to light for up to 2 hours after respiratory arrest. Death is dramatic. Overall, these are neurotoxic reactions.

What Is Contained in Snake Venom?

From the outside, the venoms of all snakes appear similar. They are slightly opalescent fluids of lemon-yellow color. However, the two types of venom (vasotoxic and neurotoxic) differ significantly in their chemical composition. They are similar only in that both are complex mixtures of biologically active substances with protein character. The venom of vipers and rattlesnakes is rich in hydrolytic enzymes, while the venom of cobras and sea snakes is rich in peptide neurotoxins.

The enzymes in the venom of Viperidae (vipers, adders) and rattlesnakes resemble those of the digestive tract. Since their venom glands are modified salivary glands, their secretion probably once served to break down animal food. Some researchers believe that the same enzymes are also found in the saliva of non-venomous snakes. However, these are not poisonous because they lack a mechanism to introduce the venom into the victim's body. The venom of vipers and rattlesnakes contains esterases, phospholipases, phosphodiesterases, nucleases, cholinesterases, oxidoreductases, hyaluronidase, and various proteolytic enzymes, including exo- and endopeptidases, bradykinase, and enzymes involved in blood formation. Phospholipases break down phospholipids. These penetrate cell membranes and destroy them. Hyaluronidase, in turn, breaks down hyaluronic acid, a component of the intercellular substance. This causes the tissue to become spongier and the walls of blood vessels to become more permeable. Some researchers believe that proteases are more responsible for the permeability of blood vessels than hyaluronidase. However, this is disputed. Proteolytic enzymes hydrolyze peptide bonds and are generally not very substrate-specific. However,

I. G. Ivanov, *The Invisible Language of Nature*,
https://doi.org/10.1007/978-3-662-73302-8_35

there are also highly specific proteases in snake venoms that act with remarkable selectivity. These include, for example, bradykinases, which break down bradykinin. This is a peptide consisting of 9 amino acids and is involved in blood pressure regulation in mammals. Fibrinolytic enzymes have also been found that dissolve fibrin and influence blood coagulation. We have already established that the blood-coagulating effect of the venom is characteristic only for the initial stages after the bite. This is not without biological significance. The rapid formation of a thrombus at the site of injury prevents the outflowing blood from carrying away the venom. This ensures complete absorption into the body.

In contrast to the venoms of vipers, the neurotoxic venoms of cobras and sea snakes have no proteolytic effect. Their active substances are low-molecular-weight polypeptides (neurotoxins) that inhibit the transmission of nerve impulses from neuron to muscle and, to a lesser extent, from neuron to neuron. To clarify the molecular mechanisms of neurotoxins, we will briefly discuss the mechanisms of nerve impulse transmission from neuron to neuron or from neuron to muscle cell.

This occurs via synapses. The synapse is a microscopic structure at the end of the neuron, consisting of three main components: the presynaptic membrane, the postsynaptic membrane, and the synaptic cleft (Fig. 1). The nerve ending, which terminates in a presynaptic membrane, is essentially a neurosecretory apparatus that produces and secretes chemical substances called mediators. In the peripheral nerves of higher organisms, this is acetylcholine. In the resting, non-excited state of the neuron, acetylcholine accumulates in small vesicles (30–50 nm in diameter) located near the presynaptic membrane. After the neuron is depolarized, the vesicles burst and release their contents into the synaptic cleft, from where the mediator diffuses to the postsynaptic membrane. There, specific receptors bind the mediator, causing the postsynaptic membrane to become polarized and the nerve impulse to be transmitted to the next neuron or effector organ. After the impulse is transmitted, the receptors are "cleaned" by the enzyme acetylcholinesterase, which breaks down acetylcholine into choline and acetic acid. Thus, the receptor is now ready to bind a new acetylcholine molecule and generate a new impulse. The mechanism of action of snake neurotoxins consists in interrupting the nerve impulse at the synapse. Depending on which stage of the transmission cascade is affected, these neurotoxins have either a curare-like or other effects. The former are found only in the venom of cobras and sea snakes, the latter in other venomous snakes.

What does curare-like effect mean? Curare refers to a group of poisons derived from plants of the genus Strychnos. Indigenous peoples of

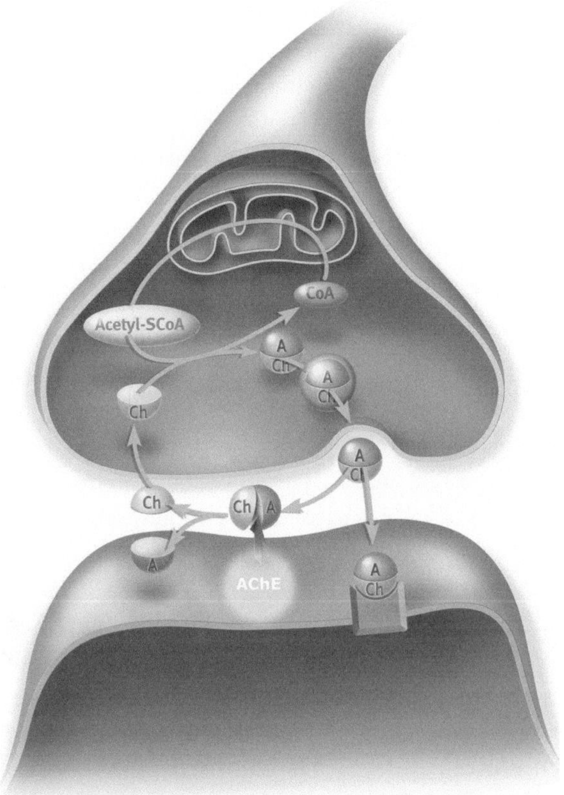

Fig. 1 Structure of the cholinergic synapse and mechanism of nerve impulse transmission (*A:* acetate [acetic acid]; *ACh:* acetylcholine; *AChE:* acetylcholinesterase). (© RFBSIP/stock.adobe.com)

the Amazon and Orinoco basins used it to coat their arrow tips. These are typical nerve poisons, consisting of alkaloids, and are still among the most potent natural toxins known. Once in the bloodstream, they block the transmission of nerve impulses from neuron to muscle, causing paralysis of the motor muscles without affecting the central nervous system. The affected person therefore dies of respiratory arrest without losing consciousness. Studies on the effects of the curare group show that the toxins have a high affinity for acetylcholine receptors and thus compete with acetylcholine. The curare alkaloids are broken down relatively quickly in animal tissue. If the victim is artificially ventilated, their effect is overcome after some time and breathing is restored. The "curare-like" effect of snake neurotoxins thus means that they act on the acetylcholine receptor. In contrast to curare toxins, however, they have a much higher affinity and form much stronger bonds with the receptor. Hence their higher toxicity. For example, while the intravenous lethal dose of D-tubocurarine (a curare-like alkaloid) in mice is 200 μg/kg, for cobra neurotoxin it is 50–100 μg/kg.

Curare-like snake neurotoxins are low-molecular-weight polypeptides, which are divided into small and large peptides according to their molecular weight. Small neurotoxins consist of 60–62 amino acid residues (corresponding to a molecular mass of about 7000 daltons), contain 4 disulfide bridges, and retain their activity even after heating to 100 °C. However, they are sensitive to strong bases, oxidizing agents, and heavy metal ions, which destroy the disulfide bridges. Since small and large neuropeptides differ in their amino acid composition and primary structure, the antiserum against one is not active against the other. Both groups of toxins have different affinities for acetylcholine receptors. The large neurotoxins bind more strongly. If a neuromuscular blockade is caused by a small neurotoxin, it can be easily reversed by flushing with neostigmine. However, this measure is unsuccessful with large neurotoxins.

If one compares the chemical formulas of acetylcholine and the neurotoxins, it becomes clear that there are no similarities between them. This logically raises the question: If acetylcholine receptors are highly specific structures that evolved in higher organisms, how is it possible that compounds with completely different chemical structures produce the same biological effect? The answer is: acetylcholine interacts with the receptor via its positively charged quaternary ammonium group (derived from choline) and a carbonyl group (from the acetate residue). These groups are also found in neurotoxin molecules. The positively charged ammonium group is mimicked by the guanidyl group of the amino acid arginine at position 37 in the polypeptide chain of the large toxins. The carbonyl group comes from the amino acid asparagine at positions 31 and 67. Both groups serve as recognition sites and for the initial binding of the toxin to the receptor, while the binding itself involves much larger areas of the molecule. Sea snake venom is extremely rich in neurotoxins. The content can be up to 85%, whereas the venom of terrestrial snakes contains significantly less. The venom of the Siamese spitting cobra *Naja siamensis* contains only 20–30% neurotoxins.

In addition to toxins with curare-like effects, snake venoms also contain other neurotoxic substances with different mechanisms of action. These include the neurotoxin of the South American rattlesnake (*Crotalus durissus terrificus*, Fig. 2). The neurotoxin has phospholipase activity and was isolated in pure form in 1956, known as crotamine. It is characterized by a molecular mass of 13,000 daltons and a lethal dose of 50 μg/kg in mice. Its effect

Fig. 2 South American rattlesnake (*Crotalus durissus terrificus*). (© AlizadaStudios/Getty Images/iStock)

consists in inhibiting the release of acetylcholine at the synapses, thereby interrupting the transmission of nerve impulses. Neurotoxins with similar effects have been found in the venom of other snakes. One of these venoms, taipoxin, isolated from the Australian taipan (*Oxyuranus scutellatus*), is the most toxic neurotoxin known to date. Its lethal dose is only 2 μg/kg in mice. In general, presynaptic neurotoxins have a higher molecular weight and are more toxic than curare-like substances.

Finally, there is another group of neurotoxins that exert their effect at the intermediate stage of nerve impulse transmission. These are the cholinesterases. They impede the transmission of the nerve impulse through the synapse by breaking down acetylcholine.

When considering the mechanisms of action of snake venoms, one cannot help but admire the adaptability of living organisms. In this context, we may ask: How does the snake know so much about the biochemistry and physiology of its victims that it can produce such perfect venoms? The point is that the snake never possessed such knowledge, nor could it control the activities of its chemical laboratory. Perfection was achieved blindly through countless random experiments, of which the most successful were confirmed by natural selection. Knowledge of the molecular mechanisms of action of snake venoms has contributed to the development of new methods for treating snakebite victims. Old methods, such as cauterizing the bite site with a hot iron, making deep cuts with a knife to let the venom drain, pouring hot oil, treatment with strong acids, bases, or oxidizing agents (potassium permanganate), drinking alcohol, etc., have long since been abandoned. Modern medicine recommends the use of the gentlest possible means, such as antivenoms. These are obtained from the blood of animals (usually horses) immunized with snake venom. The gamma globulin fraction, which contains antibodies against the proteins of the snake venom, is extracted. The antibodies bind to the toxins and convert them into inactive protein complexes, which are destroyed by specialized blood cells. An antivenom against viper venom is also produced in Bulgaria. To compensate for blood loss, blood transfusions and infusions of fibrinogen-enriched preparations are used.

We have already said that the snake is not only an enemy, but can also be useful to humans. Interest in the healing properties of snake venom began in the early 20th century, after reports emerged that epilepsy patients recovered after a rattlesnake bite. Studies actually showed that snake venom reduced seizures and alleviated the suffering of epilepsy patients. It was also observed that some chronic pain disappeared after a snakebite. This was also experimentally confirmed. Cobra venom is notable for such an effect. In low concentrations, it has an analgesic effect and can be used in cancer patients instead of morphine. In this case, it even has the advantage that its effect lasts longer and does not cause habituation. A number of pharmaceutical preparations are made from snake venom, such as Vipratox and Lebetox (Russia), Viperalgin (Czech Republic), etc., for the treatment of polyarthritis, rheumatism, sciatica, and other neuralgias.

Large quantities of venom are required for the production of antivenom and pharmaceutical preparations based on snake venom. It is obtained from snakes bred in special snake farms (serpentariums). The first such farm was established in 1899 in São Paulo, Brazil. Later, one of the world's largest institutes for the study of snakes, the Instituto Butantã, was founded there. In 1939, a large serpentarium was also opened in Tashkent, which has been expanded and renovated several times. Today, serpentariums exist in many countries, especially where snakes occur naturally.

How is snake venom collected? In the past, the snake was decapitated and the venom glands were squeezed out. Today, the venom is collected without killing the snakes. The technique consists of placing the snake on a table, grasping it behind the head, and placing a glass beaker over its mouth. When it opens its mouth, the rim of the glass is slid under the two fangs. Its head is pressed down and the venom glands are massaged from the neck to the head with the fingers (Fig. 3). The massage is considered the most delicate part of the procedure. Since 1961, snake venom has been obtained using electric current. Two electrodes are placed on the oral mucosa of the snake and a weak current (6 V) is applied. This shortens the time required

Fig. 3 Collection of snake venom. (© Ratchapong/Stock.adobe.com)

for venom extraction threefold and reduces trauma to the animal. While an experienced collector can process up to 20 snakes per hour using the classical method, with electricity he can handle 60. After the venom is collected, it is freeze-dried. This allows the active substance to be preserved for up to 50 years.

The Venom of Living Fossils—Scorpions

Scorpions are living fossils. They are the oldest terrestrial animals on Earth. They appeared in the Paleozoic era, survived all the planet's catastrophes, and have reached the present day almost unchanged. They are remarkably resilient. Their representatives live both in deserts and at altitudes of 7,500 m. Scorpions can survive up to 386 days without water and food and endure exposure to 150,000 roentgens. For humans, 600 roentgens is lethal. A macabre joke among some scientists is that after a nuclear war, only scorpions and blue-green algae will remain on Earth.

For people from countries with temperate climates, scorpions are primarily known as a zodiac sign. For those living in the tropics and subtropics, they are a real scourge. In Egypt, about 36,000 people are stung annually, and in Mexico, the number is 20,000. As a curiosity, it is worth mentioning that during the same period, the number of people bitten by venomous snakes in these countries was 2,000 and 200, respectively.

Approximately 2,000 species of scorpions live on Earth, which are divided into about 13 families based on morphological characteristics. Depending on their habitat, scorpions are classified as xerophilic (drought-loving) and hydrophilic (moisture-loving) species. The former inhabit arid and desert regions, while the latter live in damp, dark caves, under stones, and tree roots. Hydrophilic scorpions are also found in Europe. They are small, up to 5–6 cm long, and inhabit, for example, areas of the Balkan Mountains, Sredna Gora, and Belasitsa. A characteristic feature of these scorpions is that they seek shelter in wet weather and can also settle in human dwellings. It is good to know that they like to hide in shoes, bags, sleeves, and bedding.

I. G. Ivanov, *The Invisible Language of Nature*,
https://doi.org/10.1007/978-3-662-73302-8_36

Fig. 1 Black scorpion *Heterometrus.* (© 43035245/Getty Images/iStock)

The body of scorpions consists of loosely connected chitin segments and ends with a bent segment at the tail (Fig. 1). It has four pairs of legs, each ending in two claws, and a pair of pincers resembling those of crustaceans. The venom gland is located in the last segment and consists of two bean-shaped glands enclosed in a common sac. This ends in a hollow stinger for wounding and injecting venom. The venom sac is protected by a chitinous shell, and inside is a ring of muscle fibers made of hollow fibers to press the venom through the stinger into the injured tissue. The scorpion generally uses its venomous weapon for defense. Sometimes, however, it also resorts to it to kill larger insects that it cannot overpower with its pincers alone. It is believed that scorpions attack humans only in extreme emergencies. However, the numbers mentioned above show that in countries where they occur, they pose a real danger.

How does scorpion venom work? It has both hemotropic (acting on the blood) and neurotropic (acting on the nervous system) effects, with the two effects varying in intensity among different scorpion species. At the moment of the sting, a sharp pain usually occurs, as if pricked with a hot needle. Occasionally, blood may flow from the wound. The skin around the sting site quickly becomes red and swollen. After 1–2 hours, acute local pain begins, which can be very severe. Palpitations, choking fits, breathing difficulties, and headaches occur. The painful and inflammatory reactions reach their maximum after about 6–8 hours. A skin rash may also appear.

The neurotropic effect of scorpion venom unfolds via the central nervous system. Irritation of the responsive areas of the brain can lead to convulsions of individual muscle groups. The fingers, for example, cramp so severely that they can hardly be opened even by another person. In the most severe cases, paralysis of a limb occurs. The cramps can also affect the chest, tongue, and pharyngeal muscles, making swallowing and breathing difficult. The victims are highly agitated and restless, and trembling of the eyelids and fingers often occurs. In rare cases, symptoms of psychiatric disorders are observed. Some researchers compare the first phase of the neurotropic effect of scorpion venom to that of strychnine and the second phase to that of curare.

Scorpion venom can cause severe damage to both the blood and internal organs. It acts hemolytically on the blood (destroying red blood cells), and the liver and kidneys can be damaged. Neurotoxins reach the central nervous system via the bloodstream, while hemotropic substances must pass through three successive biological barriers with the aim of neutralizing their effect. The first is the site of the sting itself. Through the adjacent tissues, the hemotropic substances reach the lymph nodes (second barrier) and finally the liver (third barrier), where they are chemically altered and broken down.

The neurotoxins of scorpions are similar in their chemical composition to those of the Elapidae. They consist of single-chain polypeptides with a molecular mass of 7,000 daltons (63–65 amino acids), stabilized by four disulfide bridges. In the venom of scorpions of the genus *Centruroides*, eight such peptides have been found, and in the genus *Tityus*, six. The amino acid sequence of these neurotoxins has been determined. It has been found that they are rich in aromatic amino acids and especially in tyrosine.

Although the similarity in the molecular structure of snake and scorpion neurotoxins is great, their mechanism of action is different. Unlike snake neurotoxins, scorpion neurotoxins have no affinity for acetylcholine receptors. They cause depolarization of neuronal and muscle membranes by increasing their permeability to sodium ions. They are therefore more toxic than most snake neurotoxins. The lethal dose here is between 9 and 90 μg/kg in mice. The main protein components of the venom of the South Indian hydrophilic scorpion *Heterometrus scaber* (Black Thai Scorpion) are acid phosphatase, hyaluronidase, acetylcholinesterase, and ribonuclease (an RNA-destroying enzyme), making it similar to viper venom. The venoms of scorpions and snakes are also similar immunologically. As early as the beginning of the 20th century, it was shown that an antiserum against cobra venom also has a neutralizing effect on scorpion venom. Today, this serum is used for both cobra bites and scorpion stings.

In addition to neurotoxins, scorpion venom also contains insect toxins. These are substances that have a strongly paralyzing effect on insects but are harmless to mammals. Such a toxin was isolated from the venom of the Central Asian scorpion *Mesobuthus eupeus*. It is a polypeptide consisting of 36 amino acids, whose molecule is stabilized by four disulfide bridges.

The Chemical Weapon of Spiders

Apart from the hexagonal honeycomb cells of honeybees, there are hardly any more perfect architectural creations in nature than spider webs. Spiders weave them with such speed and virtuosity that they rightly aroused the envy of the Greek goddess Pallas Athena. According to legend, spiders originated as follows: In the region of Lydia lived a peasant girl named Arachne. She was famous for being the best weaver in all of Hellas. Her fabric was as delicate as mist and as transparent as air. She was so convinced of her abilities that she did not hesitate to compete even with Athena, the goddess of art, crafts, and handiwork, and the favorite daughter of Zeus. The contest was observed and judged by 12 gods under the leadership of Zeus. The "jury" awarded first place, as expected, to Pallas Athena. Arachne could not bear this humiliation and tried to hang herself. But the magnanimous Athena freed her from the noose and said: "Live, rebellious maiden! But from now on you shall hang and weave as long as you live." Then she sprinkled her with the juice of magical plants, her hair fell out, she shrank, and was transformed into a spider. Since then, the spider Arachne has hung in her web and continues to weave it. The large class of arachnids, Arachnida, is named after Arachne.

There are more than 52,000 species of spiders on Earth, unevenly distributed across countries and continents. For example—minimum values in each case—there are 2,494 species in Brazil, 1,346 in France, 688 in Switzerland, 556 in Great Britain, 341 in Norway, and so on. Every year, the family of spiders is expanded by 200–250 newly discovered species. Apart from their great diversity, arachnids are astonishing for their high population

I. G. Ivanov, *The Invisible Language of Nature*,
https://doi.org/10.1007/978-3-662-73302-8_37

density. About 5,000,000 individuals live on 1 hectare of forest clearing, and on the same area of tropical forest, 25,000,000.

Although spiders look different, their body structure is always the same. It consists of two parts—the cephalothorax and the abdomen. The cephalothorax bears four pairs of legs (in contrast to insects, which have three pairs), two pedipalps, and two chelicerae (jaws). The chelicerae are located near the mouth opening and are symmetrical, hollow, sickle-shaped fangs. In almost all spiders, they serve as organs of injury. In spiders, they are also used to inject venom into the body of the victim. Spiders range in size from 1 mm to more than 20 cm. Accordingly, the size of the chelicerae varies, reaching up to 2 cm in large spiders.

Spiders primarily use their chemical weapons to attack and kill their prey. Since they feed mainly on insects and other arthropods, their venom is much more potent in cold-blooded animals than in warm-blooded ones. Of the many spider species, some are poisonous to humans. One that has acquired a notorious reputation as a killer of humans in the past is the tarantula (Fig. 1). This is its common name and a collective term that includes many members of the family Theraphosidae.

Tarantulas are heat-loving animals that live in warm regions on all continents. They are dark brown to black. The body is pear-shaped, the head shield-like. Tarantulas are nocturnal, so encounters with humans are accidental. The American traveler V. Nadson wrote in 1896: "The tarantula is

Fig. 1 Tarantula. (© MirekKijewski/Getty Images/iStock)

extremely active, mobile, and irritable. When you observe it, you can't help but think that nature created it to disturb the balance. You only have to walk past its burrow, even if you are 3–4 m away, and it will leap out and chase you for 30–40 m."

100–200 years ago, some peoples experienced a real psychosis regarding tarantulas (Italian: Tarantola). In southern Italy, for example, the tarantula was considered a deadly enemy of humans. It was believed that a person bitten could only be freed from its venom by profuse sweating. Therefore, the bitten were made to move in a rapid rhythm until they sweated to the sounds of a special melody called the Tarantella.

Far more dangerous to humans is another species of spider—*Latrodectus tredecimguttatus* (Fig. 2). Here it is called the black widow, in Arab countries Karakurt (black wolf). It reaches a length of 12–15 mm. The spider has a velvety red color and a large, egg-shaped abdomen with 13 dot-like grooves, which are not always visible. It is found in almost all of Europe, Asia, North and West Africa. The spider is common, for example, along the Black Sea coast, the Balkan Mountains, and western Turkey.

It lives mainly in meadows and fields and builds its nests close to the ground, often using abandoned rodent and insect burrows. Unpleasant encounters with the black widow usually occur during fieldwork, especially when gathering hay or during outdoor recreation. According to Bulgarian doctors, the consequences of poisoning by the black widow can be more severe than those caused by tarantulas. The bite resembles a sudden sting. The first phase is asymptomatic. However, after about half an hour, pain appears in the lower back, chest, and especially the abdomen. Chills set in,

Fig. 2 Black widow (*Latrodectus tredecimguttatus*). (© Frank Buchter/Getty Images/iStock)

and patients enter a strange psychological state with severe nervous agitation, restlessness, and a feeling of impending death. This is intensified by the onset of pseudo-paralysis of the lower limbs. Body temperature rises. The pronounced clinical picture lasts 24–36 hours, after which improvement occurs. The literature describes fatalities. For example, in California, 32 out of 578 bitten individuals died, and in North Carolina, 32 out of 34. It is estimated that the mortality rate from the bite averages 2–4%.

What is the reason for the high toxicity of the black widow? Its venom is rich in neurotoxins. Several proteins with a molecular weight of about 130,000 daltons have been isolated from it, which are capable of blocking neuromuscular impulse transmission. Unlike the neurotoxins of snakes and scorpions, the mechanism of action is different. Acetylcholinesterase is inhibited. As a result, an excess of acetylcholine accumulates in the synapses, preventing the synaptic membranes from restoring their electrical potential and transmitting new impulses. The neurotoxins of the black widow act similarly to nerve-paralyzing organophosphorus compounds.

We have seen that the toxins of this spider and the tarantula, despite their different mechanisms of action, are protein in nature. However, this is by no means true for all arachnids. The chemical weapons of some arachnids are based on low-molecular-weight compounds. For example, the defensive secretion of harvestmen (order Opiliones) contains the ketones 4-methylheptanone-3,4,6-dimethylnonen-6-one-3 and 4,6-dimethylketone-6-one-3. Some resemble the alarm pheromones of ants. The venom of harvestmen of the family Gonyleptidae, known as gonyleptidin, is a mixture of 2,3-dimethylbenzoquinone and 2,3,5-trimethylbenzoquinone.

Just as there are spitting snakes, there are also spitting arachnids. One such is the tropical whip scorpion *Mastigoproctus giganteus*, 2–5 cm long. It can shoot a jet of caustic liquid up to 80 cm. Chemical analysis shows that it consists of acetic acid (84%) and caprylic acid (5%).

With a Hint of Almonds

People often learn about the smell of bitter almonds from romance and crime novels before ever encountering a bitter almond themselves. This scent is usually associated with one of the most well-known chemical poisons, potassium cyanide, also known as cyanide. However, dry potassium cyanide smells no more like bitter almonds than table salt does. The accompanying bitter almond odor comes from hydrogen cyanide, which is released during hydrolysis. This, in turn, requires water or moisture. Both potassium cyanide and hydrogen cyanide are potent cellular poisons. Due to the high affinity of the cyanide anion for iron, it binds tightly to hemoglobin and iron-containing enzymes, thereby blocking redox processes in the cell. The cyanide anion is highly toxic to both lower and higher organisms.

Against the backdrop of all that is known about hydrogen cyanide, the report by an American research group that certain species of millipedes produce hydrogen cyanide was a sensation at the time. In studies with the millipede *Apheloria corrugata* (Fig. 1), the American scientists noticed that when threatened, this millipede emitted a smell of bitter almonds. Bitter almond odor is also found in other animals, without hydrogen cyanide being involved. These are other organic compounds that are completely harmless to animals and humans. To test whether the secretions released by the millipede were toxic, the researchers placed the millipede in a closed container with ants and other insects. After some time, they found that all except the millipedes were dead. This prompted the scientists to intensify their investigation of the millipede. The first task was to clarify the chemical nature of the poisonous gas. Using a gas chromatograph and qualitative reactions, it

I. G. Ivanov, *The Invisible Language of Nature*,
https://doi.org/10.1007/978-3-662-73302-8_38

Fig. 1 Millipede (*Apheloria corrugata*). (© tbradford/Getty Images/iStock)

was confirmed that the gas smelling of bitter almonds was indeed hydrogen cyanide. The following questions arose: In what form is the poisonous gas present in the body of the millipede? Which organs produce it? Why is the millipede resistant to it?

No specific anatomical structures for storing hydrogen cyanide were found in the body of the millipede. However, peculiar glands were discovered, symmetrically arranged on each segment near the legs. When these glands are stimulated, they secrete a liquid with the smell of bitter almonds. Each of these glands consisted of two sectors, which the researchers referred to as the storage chamber and the antechamber

Neither hydrogen cyanide was found in the storage chamber nor in the antechamber. It was present only in the droplet of liquid secreted when the feet were stimulated. Thus, hydrogen cyanide was not present in finished form, but was produced only at the moment of the defensive reaction. The question is: What is the nature of the precursor? To answer this question, the scientists examined the chemical composition of the gland contents. They found benzaldehyde, a compound that can bind hydrogen cyanide in the form of cyanohydrin. However, since cyanohydrin is unstable, it is stored in nature in the form of a glycoside, i.e., bound to a carbohydrate.

Chemically, this is amygdalin (Fig. 2). It is found in the seeds of bitter almonds, cherries, sour cherries, plums, peaches, and other fruits. Upon hydrolysis, glucose, benzaldehyde (also with the smell of bitter almonds), and hydrogen cyanide are produced. In the body of the millipede, amygdalin is secreted into the "storage chamber" and broken down by specific hydrolytic enzymes, such as those found in the "antechamber" of the poison glands.

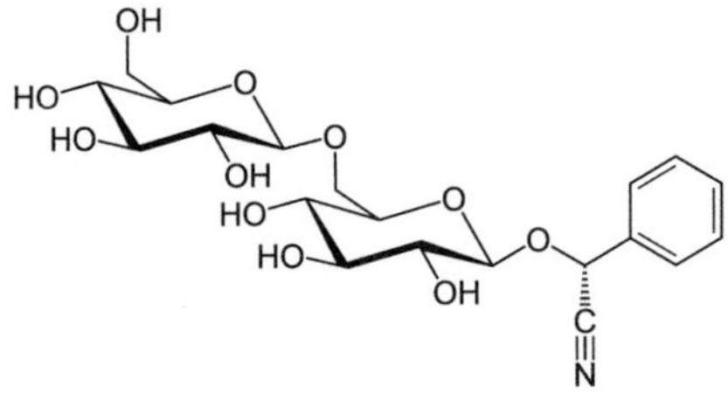

Fig. 2 Amygdalin

Based on this information, the function of the millipede's poisonous weapon can be surmised. When stimulated, the valve separating the two chambers of the poison gland opens. The amygdalin moves from the storage chamber into the antechamber and, after mixing with the enzymes present there, breaks down into hydrogen cyanide, benzaldehyde, and glucose. The reaction is initially intense but gradually subsides after the amygdalin is depleted, lasting only a few minutes. The amount of hydrogen cyanide produced by a single millipede is 0.55 mg. This is enough to kill a mouse. However, the millipede's secretion is not lethal even to small animals, as hydrogen cyanide is volatile and cannot reach lethal concentrations in open air. Nevertheless, it is an effective means against insects and insectivorous animals. When the millipede encounters an anthill, it is quickly attacked by the voracious and omnivorous ants, but it curls up in a spiral and deploys its chemical weapon. Moments later, the ants scurry away, intoxicated by the cyanide cloud. Those that manage to overpower the millipede sink their jaws into its body and clean it for a long time, as if trying to remove something disgusting. Interestingly, the ants attack the millipede again 20–30 minutes after the first assault, i.e., when the hydrogen cyanide has already evaporated. The reason for their behavior is that not only is hydrogen cyanide unpleasant for the ants, but so is benzaldehyde. It is less volatile and remains longer on the millipede's body. Thus, it has a double chemical defense—hydrogen cyanide and benzaldehyde. Without this protection, the millipede would be very vulnerable, as its hydrogen cyanide reserves would last no longer than 2–3 minutes. The second defense, however, provides at least 20–30 minutes of security. That is enough time to escape to safety. The millipede's chemical weapon also protects it from larger animals such as frogs, lizards, and birds.

Apheloria corrugata is only one representative of the large class of millipedes (Myriapoda). This class consists of the subclasses Chilopoda and Diplopoda. The former have well-developed organs for wounding and injecting venom, while the latter, to which *Apheloria* belongs, lack such tools. They spread the toxic secretion over the surface of their bodies. This results in a different purpose for the venom weapon in the two millipede groups. For the former, it serves as a means of attack and hunting, while for the latter, it is a means of defense. Evolution has led *Apheloria* to use its venom weapon economically. When irritated, it does not activate all its poison glands, but only those near the irritated area. If only one of its legs is pinched with tweezers, two glands are involved. Only in cases of very strong and prolonged irritation are all glands triggered as an exception.

In contrast to the Chilopoda, Diplopoda use other chemical compounds for protection. Some secrete quinones and cause redness of the skin when touched with bare hands, others secrete paracresol (a repellent against insects and small predators). Some smell of camphor. The poison glands of these millipedes resemble those of *Apheloria* with the difference that they do not have "economization." In their case, this is not necessary, as quinones and phenols are stable and only weakly volatile compounds and can be stored as "finished products" in the respective storage organs.

The Chemical Weapon of Hymenoptera

The Hymenoptera are the largest order of insects. They originated in the Triassic and include over 150,000 living and 2,000 extinct species. Their name derives from the membranous appearance of their wings, which resemble a thin, transparent zipper. This order includes the families of bees (Apoidea), wasps (Vespidae), ants (Formicidae), and others. Most of these insects sting and possess a well-developed system for producing and injecting venom into the body of a victim (Fig. 1).

Anatomically, the stinging apparatus of bees and wasps is the best studied. The bee's stinger is a modified ovipositor, present only in female individuals. Thus, only the queen and worker bees possess a stinger. As a rule, the queen uses her stinger only in fights with other queens.

The stinger is located at the end of the bee's body, on the rearmost segment, and consists of two hollow, barbed needles (stylets). With the help of a strong muscle, the needles are driven into the victim's body. The barbs on them ensure unidirectional movement and prevent withdrawal (Fig. 2).

As a result, the bee's stinger remains at the site of the sting—at least in vertebrates—and vital internal organs are torn out. This leads to the bee's death 1–2 hours after the sting. In other words, the bee commits suicide with its sting. Nature has been kinder to wasps in this respect. Their stylets are smooth and can be easily withdrawn from the victim's tissue, allowing a wasp to sting repeatedly.

At the base of the stinger are a large and a small gland. The large gland produces the venom. The secretion of the small gland serves to activate the venom produced by the large gland. The secretions of both glands collect in

I. G. Ivanov, *The Invisible Language of Nature*,
https://doi.org/10.1007/978-3-662-73302-8_39

Fig. 1 Honeybee (left) and wasp (right). (© schnuddel/Getty Images/iStock)

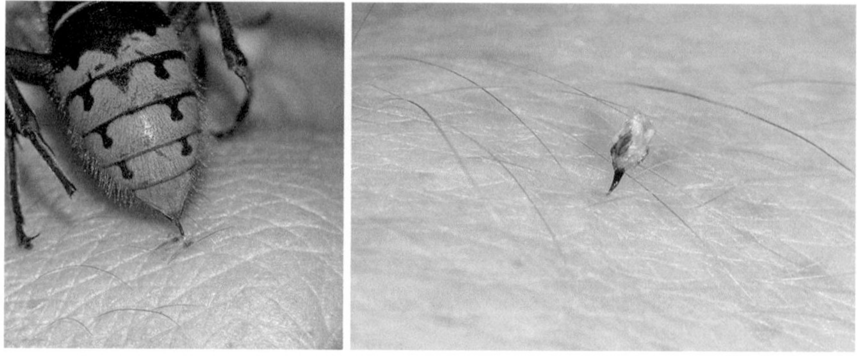

Fig. 2 Stinger of a wasp and stinger of a bee. (left: © ChriSes/Fotolia, right: © mirkograul/Stock.adobe.com)

a pyramid-shaped sac that is connected to the hollow space of the stinger. The amount of toxic secretion depends on the age of the bee. The most productive bees are 17–18 days old. The sting is accompanied by acute pain. This is not due to the sting itself, but to the injected venom. After the sting, the stinger continues to move under the influence of the muscles attached to it, causing it to penetrate deeper.

Animals vary in their sensitivity to bee venom. The bee itself is the most sensitive. For a bee, the simultaneous sting of several bees can be fatal. Humans are generally less sensitive to bee venom. Moreover, for humans it can also be a remedy. Since ancient times, bee venom has been used to treat rheumatism, gout, sciatica, arthritis, myositis, and so on.

What Do Hymenoptera Venoms Contain?

The venoms of hymenopterans are complex mixtures of various organic compounds, proteins, and peptides that can cause local pain, inflammation, itching, irritation, and moderate to severe allergic reactions. Of all these venoms, that of the honeybee *Apis mellifera* is the best studied. It is rich in enzymes and peptides, but also contains histamine, serotonin, dopamine, noradrenaline, and polyamines. Among the enzymes, phospholipase A2 (responsible for the breakdown of membrane phospholipids), hyaluronidase (breakdown of hyaluronic acid found in the intercellular matrix), acid phosphatase (removal of phosphate groups from organic phosphates), etc., are notable. Among the peptides, melittin (a lytic polypeptide), apamin (a neurotoxic peptide), and mast cell degranulating peptide (MCD) are well known. The bee venom enzymes are proteins with a molecular mass between 15.0 and 50.0 kDa. Phospholipase A2 (15.0–16.0 kDa) accounts for about 12% and is the largest component. In addition to causing edema, it also has cardiotoxic, neurotoxic, myotoxic, anticoagulant, and other effects, and is a strong allergen. Phospholipase breaks down the phospholipids of cell membranes into lysophospholipids and long-chain fatty acids, creating pores in the cell membrane and thereby destroying the cell. Numerous studies have demonstrated a synergistic reaction between phospholipase A2 and melittin (see below). The result is hemolysis (destruction of red blood cells). Hyaluronidase is an enzyme responsible for the breakdown of hyaluronic acid and chondroitin sulfate, substances found in connective tissue and intercellular tissue. Hyaluronidase makes the environment of the stung tissue more spongy and more accessible to the venom. It is therefore also

I. G. Ivanov, *The Invisible Language of Nature*,
https://doi.org/10.1007/978-3-662-73302-8_40

referred to as a "venom spreading factor." Its molecular weight ranges from 30.0–60.0 kDa. Acid phosphatase or phosphomonoesterase makes up about 1% of the dry matter of bee venom. It removes phosphate groups from various organic substrates and is active at low (acidic) pH values. Other enzymes found in bee venom as well as in the venoms of other hymenopterans include alpha-glycosidase (0.6% of the dry weight), lysophospholipase, various esterases, lipases, etc. Lipases are thought to be involved in the processes of cell lysis (disintegration of a cell). The venoms of hymenopterans also contain proteases. It is suspected that they cause moderate necrosis. The best studied is the protease of the bumblebee (*Bombus*), which is responsible for some severe allergic reactions. These arise from the interaction of mast cells with immunoglobulin E (IgE). A cascade of mediators is triggered, including histamine, leukotrienes, platelet-activating factors, enzymes, peptides, etc., which cause persistent local inflammation. In rare cases, this reaction can lead to cardiorespiratory depression, systemic anaphylactic shock, and death. In most cases, stings from hymenopterans are mild. Typically, they are accompanied by pain, local inflammation, and itching, which subside after a few hours.

As already mentioned, the venoms of hymenopterans are peptides. The best studied of these venoms is melittin. It makes up 40–50% of the dry mass of bee venom. It has alkaline properties and appears to be the main component responsible for causing local pain. Melittin consists of 26 amino acids with amphipathic (polar and nonpolar) properties, which enable it to interact with phospholipids and thus increase the permeability of cell membranes. Its tetrahedral structure is believed to possess ionophoric properties, which are responsible for the persistent depolarization of skin neurons and are thought to underlie the pain of a sting. Melittin also induces lysis in many cell types, accompanied by the release of cellular enzymes and other biologically active substances into the intercellular space. Furthermore, melittin can cause either dilation or constriction of blood vessels, thereby affecting cardiac and skeletal muscle.

Another peptide in bee venom is apamin. It accounts for about 2% of the dry mass of the venom, consists of 18 amino acid residues (2.0 kDa), and has neurotoxic effects. In small mammals, apamin causes convulsions, but it does not have serious effects in humans. Like other neurotoxins, apamin binds to receptors on the postsynaptic membrane and thereby inhibits or hyperactivates alpha-adrenergic, cholinergic, and purinergic receptors. These effects are related to the blocking of postsynaptic ion channels, which play an important role in the efficiency of repetitive activities in neurons in both vertebrates and invertebrates.

Mast cell degranulating peptide (MCD) is also present in bee venom and is chemically similar to apamin. It consists of 22 amino acid residues, makes up 2% of the dry substance of the venom, and is likely the main factor causing the massive release of histamine after a sting. It is believed to bind to specific membrane receptors, similar to apamin.

In contrast to the venoms of bees and wasps, the venom of ants is significantly less studied. The venom of ants from the subfamilies Myrmicinae, Ponerinae, and Formicinae is the best known. Their venom contains organic acids, histamine, hyaluronidase, quinine-like substances, etc. In some cases, such as in the venom of Formicinae, neurotoxins may also be present.

The Bombardier Beetle

Hardly any popular name fits an animal as well as that of the bombardier beetle (*Brachinus crepitans*). In the truest sense of the word, it can fire shots by ejecting a hot gas mixture with a temperature of over 100 °C over a distance of 25–30 cm. Its chemical weapon acts much like a flamethrower. This beetle is capable of firing up to 70 consecutive shots at a rate of 500 shots per second. What does the bombardier beetle look like, and how does its weapon work?

About 500 species of bombardier beetles are known, belonging to the genus *Brachinus* (family Carabidae), and they mainly inhabit warm regions. Some species, such as *Brachinus crepitans* (Fig. 1), *Brachinus explodens*, and *Brachinus psophia*, can also be found in our area.

The first reports that some beetles of the genus *Brachinus* emit a gas mixture when disturbed date back to 1778. Pastor Wilhelm described the behavior of such a beetle in 1796. Its shots resembled a pistol shot, and after firing, the air had a faint smell of dust. In 1808, Dr. L. Dufault, a physician in Napoleon's army, described that during the march through the Pyrenees into Spain, yellow-brown spots of unknown origin sometimes appeared on the soldiers' skin. After careful observation, he found that the spots, which resembled burns from gas, were due to the action of a particular beetle species. In 1899, it was determined that the gas mixture reached a temperature of over 100 °C.

Today, it is known that the chemical weapon of the bombardier beetle is based on two chemical reactions, catalyzed by two different enzymes. The first is the decomposition of highly concentrated hydrogen peroxide (H_2O_2)

I. G. Ivanov, *The Invisible Language of Nature*,
https://doi.org/10.1007/978-3-662-73302-8_41

Fig. 1 Bombardier beetle (*Brachinus crepitans*) (Udo Schmidt, CC BY-SA 2.0, Wikimedia Commons)

into oxygen and water under the action of the enzyme catalase. The second is the oxidation of hydroquinone and methylhydroquinone to 1,4-benzoquinone and 1,4-methylbenzoquinone by the oxygen released from the hydrogen peroxide. The second reaction is highly exothermic and releases a large amount of heat, sufficient to heat the reaction mixture to over 100 °C and eject it over a distance of 25–30 cm.

How is the bombardier beetle's "flamethrower" constructed? It is located at its rear end and resembles a rocket, from whose nozzle a hot gas mixture is expelled (Fig. 2). As shown in the figure, the "explosion chamber" has a thick outer chitin shell, which serves both as heat protection and ensures mechanical strength. The chamber is connected to two "tanks" for hydrogen peroxide and hydroquinone, which are produced by specialized glands. The concentration of hydrogen peroxide in the tank reaches up to 25%, which is truly remarkable for chemists. They know that H_2O_2 is difficult to concentrate and store. Since the maximum possible concentration is 40% (perhydrol), it is assumed that the tank contains stabilizers, the chemical nature of which is unknown. It is only known that the tank is sealed by a valve. Equipped with a muscle, the valve is opened in case of danger to mix the reactants in the "explosion chamber." At the same time, microscopically small glands are activated that produce the enzymes catalase and peroxidase. Their mixing with the reaction mixture leads to the explosive reaction described above, in which the temperature rises to over 100 °C within a few seconds. This, in turn, leads to a sharp increase in pressure and the spontaneous ejection of the hot gas mixture.

The temperature of the gas jet was originally calculated based on thermodynamic knowledge of the chemical reactions involved. Calculations show that if the entire amount of reactants behaves according to the described equations, 1 mg of the reaction mixture releases 0.19 cal (calories) of heat, which is enough to vaporize 20% of the reaction mixture and heat it to 100 °C. To verify this theoretical result, a team of American scientists

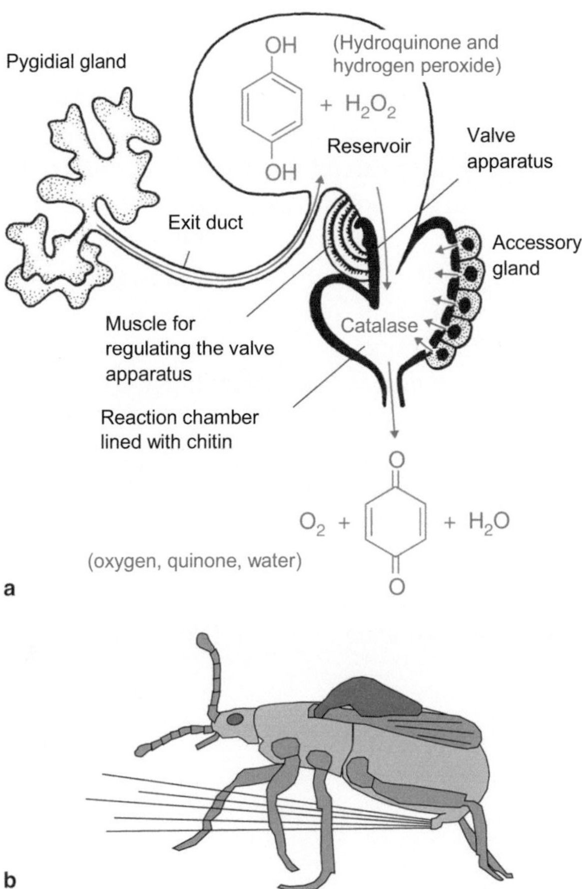

Fig. 2 Structure and function of the bombardier beetle's firing apparatus

constructed a highly sensitive microcalorimeter for direct heat measurement. It consists of a cylinder made of thin copper foil, connected to a sensitive thermoelement and a recording device. The beetle is lowered into the cylinder with a special hook, forced to fire a shot, and the released heat is measured directly. The experimentally determined average value is 0.22 cal per 1 mg, which is close to the theoretically calculated value (0.19 cal/mg). The same authors also directly measured the temperature of the gas mixture with a highly sensitive thermoelement and an oscilloscope, and determined a value of about 100 °C. Until recently, such high temperatures were considered incompatible with living matter, and the very fact of their detection is impressive.

How does the bombardier beetle avoid the harmful effects of high temperatures on its own body? The thick chitin shell lining the inner wall of the explosion chamber helps. Chitin is a good thermal insulator and exchanges heat only with difficulty with the surroundings, including adjacent soft tissues. In addition, the temperature of the gas mixture rises within 1–2 s. This short time is not enough for intensive heat exchange with the internal body fluids. Almost all the heat is carried away with the expelled gases. The chitin shell also protects the beetle's interior from the toxic effects of hydrogen peroxide and quinones. Otherwise, hydrogen peroxide would trigger reactions that could damage cell membranes, intracellular proteins, and nucleic acids. Quinones are also reactive and can denature proteins with which they come into contact.

The chemical weapon of the bombardier beetle is a reliable means of defense. It repels not only small rodents but also larger insectivorous animals and birds. Once an animal comes under bombardment from the beetle, it becomes stressed and develops a stable conditioned reflex for the rest of its life. After such an incident, it permanently excludes these insects from its diet. Some small predators are even blinded by the hot gas jet and must starve.

Do Mammals Possess Chemical Weapons?

Nature has equipped mammals with a more highly developed brain, but at the same time has "withheld" chemical aids from them. It is difficult to find an acceptable explanation for this fact, but it is obvious that a more advanced nervous system is better combined with strong teeth, sharp claws, powerful muscles, and swift feet than with venom glands. However, there is one mammal known that can be counted among the venomous animals: the platypus (Fig. 1). It has a hollow venom spur on its hind legs, which is connected to venom glands. With this, it can inflict injuries on a predator and inject venom into the body. The venom is not lethal to humans, but it is to small animals. Chemically, the venom of the platypus is little studied. Physiological studies have shown that it affects blood coagulation. It causes clots that can block important blood vessels.

In addition to the platypus, a few other mammals are known to possess chemical weapons for defense, but not for the use of venom. These are animals with well-developed anal glands that produce foul-smelling liquids in emergencies. The undisputed champion here is the American skunk (Fig. 2). It can spray a foul-smelling liquid up to 3 m (and according to some authors, up to 12 m) away. The odor can be detected hundreds of meters from the scene.

The best impression of the odor's properties is given by Gerald Durrell's description: "The dogs stood at a respectable distance, gathered around something in the grass. We shone our flashlights. There stood, defiantly, a cat-sized animal with a black-and-white tail held high. It was a black-and-white skunk. It watched us without the slightest trace of concern and was

I. G. Ivanov, *The Invisible Language of Nature*,
https://doi.org/10.1007/978-3-662-73302-8_42

Fig. 1 Platypus (*Ornithorhynchus anatinus*). (© Vac1/Getty Images/iStock)

Fig. 2 American striped skunk (*Mephitis mephitis*). (© Gerald Corsi/Getty Images/iStock)

clearly convinced that it was superior to both us and the dogs. Occasionally, it gave a short growl and made two or three little jumps in our direction. Had we approached any closer, it would have turned its back to us and looked over its shoulder in warning. The dogs, who knew very well that the skunk could spray them with its stinking liquid, kept a respectful distance. As the skunk continued to demonstrate its superiority, one of the dogs could no longer restrain itself. It lunged and tried to bite it. The skunk leapt into the air and turned so that its back was to the dog. In the next moment, the dog was rolling in the grass, whining and wiping its face with its paws. The cold night air was filled with the sharpest and most disgusting stench imaginable. Even though we were at a great distance, the stench forced us to retreat even further back. We coughed, began to gasp, and tears streamed from our eyes. It was as if we had inhaled ammonia. After

Fig. 3 Wolverine (*Gulo gulo*). (© slowmotiongli/Getty Images/iStock)

the skunk had demonstrated its power, it ran toward the dogs, jumped at them two or three times, and chased them away. Then it did the same to us. And we retreated like the dogs. It waved its beautiful tail and trotted across the meadow with an expression of satisfaction. We no longer felt like staying nearby. We called the dogs and continued on our way. The sprayed dog stank for another 3 or 4 days. The smell gradually faded."

What is the chemical composition of the skunk's foul-smelling spray? According to older sources, it contains the foul-smelling butyl mercaptan and mercaptan. These substances are known to be among the most unpleasant-smelling chemical compounds. However, this was disproved in 1974. It was shown that skunk secretion consists of crotyl mercaptan and crotyl disulfide.

The wolverine *Gulo gulo* (Fig. 3) from the family Mustelidae also emits a foul-smelling liquid. In its case, this serves to mark its territory. Since its musk glands are very active and it has a large reservoir, its musk is also used as a weapon when necessary. Similar weapons are also found in some members of the skunk family, such as the Palawan stink badger (*Mydaus marchei*) and the hog-nosed skunks (*Conepatus*).

What Animal Venoms Can Do for Humanity

Introduction

Over the millennia, humans have developed a deep-rooted fascination with venomous animals. They appear in legends and myths worldwide. A prime example of this is their significance in Egyptian mythology, where venomous animals, especially snakes and scorpions, play an important role. They were often regarded as symbols of power, protection, or danger. The cobra, a venomous snake, was a symbol of royal authority and was associated with the goddess Wadjet, who was considered the protector of the king. Scorpions were also seen as protective symbols, particularly in relation to the underworld and the afterlife. They were associated with the goddess Selket, who helped the dead protect themselves from dangers and ensured their safe passage into the afterlife. Despite their dangerous nature, these venomous animals were often revered and respected in Egyptian mythology. In contrast, in the modern Western world, venomous animals are more often met with fear and disgust.

Although fears of venomous animals are widespread, they have gained popularity in the media and can even be found as unconventional pets. This apparent contradiction between fear and attraction can be traced back to the real dangers that some of these animals, especially in tropical regions, pose to human health. Snakebites alone claim several hundred thousand lives each year. Paradoxically, the toxic cocktails produced by venomous animals

I. G. Ivanov, *The Invisible Language of Nature*,
https://doi.org/10.1007/978-3-662-73302-8_43

also possess beneficial properties. Due to their potent effects on pathologically relevant target molecules, animal toxins are promising starting points for the development of new therapeutics. Existing drugs for the treatment of cardiovascular diseases, pain, and diabetes have already been derived from these toxins. Remarkably, only a few of the world's venomous animals have been studied in detail to date. In recent years, however, a veritable global race has emerged, driven by technological breakthroughs in bioinformatics, biotechnology, and chemical analytics. Research groups around the world are competing to discover new lead structures from animal venoms that could potentially be used against diseases. This development signals a significant advance in the research and utilization of animal venoms for medical purposes.

Toxic Diversity in the Animal Kingdom

Venoms are widespread throughout the animal kingdom and occur in nearly every animal group. The proportion of venomous species is estimated at about 15%, and the evolution of toxicity has occurred convergently in various groups. This diversity extends beyond well-known venomous animal groups, including sea anemones, snails, insects, and even birds. Even within mammals, some species have developed venoms. The slow loris, for example, a nocturnal, arboreal primate native to Southeast Asia, defends itself by using venom.

The diversity of venomous animal groups is also reflected in the chemical complexity of their venoms, which often comprise hundreds to thousands of components, known as toxins. Venoms are divided into two classes, depending on how they are deployed. The first class includes passively venomous animals (referred to in English as "poisonous"), which either produce their own toxins or accumulate them from the environment, and whose toxins act upon contact or ingestion. The second class consists of actively venomous animals (referred to in English as "venomous"), which produce their own toxins and have often developed sophisticated injection systems such as fangs or stingers. Through these structures, they can inject their toxins deep into the body of their target.

The way venoms are deployed in both classes of animal toxins has significant effects on the biochemical composition of the venom cocktail. The speed at which biomolecules migrate through tissue depends greatly on their size, with smaller molecules spreading more rapidly. Passively venomous animals, which cannot inject their toxins, therefore rely heavily on the migration of their toxins through various tissue types to reach their site of action, which influences the efficiency and functionality of the venom. For this reason, the venoms of passively venomous animals consist mainly of very small organic molecules such as alkaloids, polyamines, or steroids. Thanks to their small size, these toxins quickly reach their site of action and exert their pharmacological effects rapidly, even in the absence of an injection structure. In contrast, actively venomous animals can inject their venoms directly into the organism of their opponent. In mammals, the injected toxins very quickly move from the extracellular matrix through the lymphatic system into the bloodstream and from there to their site of action. In animals without a closed circulatory system, mainly arthropods, the toxins are injected directly into the hemolymph and rapidly distributed from there. Since migration through tissue layers plays a lesser role for actively injected venoms, the influence of molecular size is less critical. Therefore, actively deployed animal venoms tend to contain macromolecules, especially proteins and peptides, whose size is several orders of magnitude greater than the components of passive animal venoms.

Animal venoms serve a variety of biological functions. The venoms of passively venomous animals are used primarily for defense against predators, pathogens, and parasites. In contrast, the venoms of actively venomous animals serve a much broader range of functions. Although they are also used for defense against predators, their most important function is prey capture. Almost every known actively venomous animal species uses venom to obtain prey. It is well documented that the specific prey spectrum of a species is a crucial evolutionary driver for the development and gradual modification of venoms. In addition to defense and prey capture, actively deployed animal venoms serve a third central function in intraspecific competition, especially in territorial fights or for reproductive success. An outstanding example of this core function is the male platypus (*Ornithorhynchus anatinus*), which uses its painfully acting venom to fight other males for mating rights. Moreover, actively deployed animal venoms also fulfill rarer secondary functions, such as intraspecific communication, immune functions, or reproduction.

Fig. 1 The highly venomous bark scorpion *Tityus serrulatus.* (© PorqueNoStudios/Getty Images/iStock)

Humans and Venomous Animals—a Contradictory Relationship

As already mentioned, the relationship between humans and venomous animals is rather ambivalent. At first glance, venomous animals and their toxin mixtures present a medical problem. Especially in the Global South, particularly in Southeast Asia and sub-Saharan Africa, snakebite envenomations are a significant medical issue. Each year, there are more than 5 million snakebites worldwide, resulting in about 2 million cases of snakebite envenomation, at least 100,000 deaths, and half a million bite victims left with permanent physical disabilities. The majority of these cases occur in the aforementioned regions of the Global South, particularly in remote and rural areas where medical infrastructure is often inadequate and therefore only a fraction of actual snakebites are recorded. It is therefore assumed that there is a significant number of unreported snakebites worldwide, and in fact, several million people may be affected by this problem. In response, the World Health Organization classified snakebites in 2017 as a neglected tropical disease of the highest category (Priority 1). Considerable efforts are being made to combat the medical consequences resulting from snakebites, and it is an important structural goal of the World Health Organization to halve the number of snakebite envenomations by 2030.

Although snakes represent the most significant health risk among venomous animals worldwide, other groups can also be of importance. The most medically significant venomous animal after venomous snakes is certainly the bee. Bees, whether domesticated or wild, are found all around the globe and often live in close proximity to humans. Bee stings therefore occur very frequently. Although the painful venom of a single bee is not particularly dangerous to humans per se, severe envenomations can occur in cases of swarm attacks, in which the victim may suffer dozens to hundreds of stings. However, such massive attacks are rather rare. There is, however, another danger inherent in bee venom. Their venoms can cause allergic reactions in humans, which can lead to anaphylactic shock and thus potentially life-threatening outcomes. It is estimated that about 3% of the world's population is allergic to bee stings, making several million people potentially at risk from bees. In some regions of North Africa, the Middle East, and Central America, scorpions also pose a major medical problem. Here, tens of thousands of people can be affected by scorpion stings, and especially children and people with pre-existing conditions can die from the resulting envenomations. The vast majority of these medically significant scorpions belong to a single family, the primitive Buthidae. Of particular relevance are the genera *Androctonus* and *Leiurus* in North Africa, *Parabuthus* in South Africa, *Hottentotta* in West Africa, the Middle East, and parts of Asia, as well as *Tityus* (Fig. 1) and *Centruroides* in Central and South America.

Globally, the often-feared spiders, on the other hand, play hardly any role, as less than 1% of all spiders are even capable of causing envenomation in humans, such as members of the genera *Latrodectus* (black widows), *Phoneutria* (wandering spiders), *Atrax* (Australian funnel-web spiders), or *Loxosceles* (recluse spiders). Since severe envenomations from spiders are rare, there are only a few spider bites with systemic effects worldwide, and fatalities are virtually unheard of.

The medical challenges posed by venomous animals stand in stark contrast to their great potential for human use. Especially in biomedicine, venomous animals and their toxins have played a significant role for about 50 years. This is due to the molecular mode of action of the toxins, which exert their destructive effects through precise interactions with selected target structures in the body. Neurotoxins, for example, interact with surgical precision with specific ion channels in the membranes of nerve cells, thereby affecting signal transmission. This can lead to paralysis or convulsions, depending on the toxin's mechanism of action. On the other hand, hemotoxins interact with components of the circulatory system, often with the blood coagulation cascade. This either impairs the blood's ability to clot or overstimulates the coagulation cascade. The result can be uncontrolled bleeding into tissues or the formation of blood clots. A multitude of animal

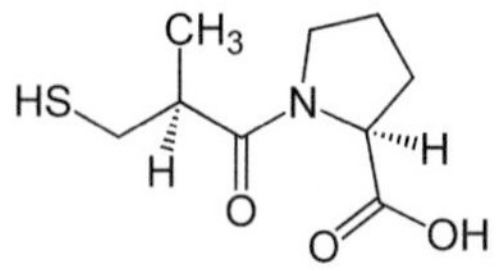

Fig. 2 Structural formula of captopril, an antihypertensive drug derived from snake venom

toxins interact with target molecules involved in the development of various diseases. In theory, these toxins can be used specifically to alleviate or eliminate disease symptoms by controlling the underlying biochemical processes.

To date, several active substances derived from animal venoms have already been isolated, further developed, and approved as therapeutics. An outstanding example is captopril (Fig. 2), probably the most important animal venom-based drug. Captopril inhibits the angiotensin-converting enzyme (ACE), a molecule that plays a crucial role in regulating blood pressure. Captopril is originally derived from teprotide, a small nonapeptide from the venom of the Brazilian jararaca pit viper *Bothrops jararaca*. Clinically, the effect of teprotide manifests as a rapid drop in blood pressure, which can lead to shock. In the 1970s, teprotide was gradually chemically modified, primarily by reducing its size. In the early 1980s, it was patented and approved as captopril (Fig. 2). Captopril is about 80% smaller than its natural prototype and therefore distributes much more rapidly in the body, but retains its inhibitory effect against ACE. It is successfully used to treat hypertension and other cardiovascular diseases and is considered the progenitor of the class of ACE inhibitors, one of the most important classes of drugs for treating blood pressure problems.

Another remarkable medication derived from animal venoms is ziconotide, obtained from the venom of a cone snail of the genus Conus. Ziconotide is a potent neurotoxin that inhibits the activity of ion channels within the pain signaling cascade and can thus relieve pain. Compared to morphine, it is about a thousand times more potent, but has virtually no addictive potential. Therefore, ziconotide is used as a powerful analgesic. Unfortunately, the use of ziconotide is relatively often associated with undesirable side effects. Moreover, the target molecules of ziconotide are located in the human brain, but the molecule itself cannot independently cross the blood-brain barrier. For this reason, ziconotide must be continuously injected into the patient via a surgically implanted pump. Therefore, the medical use of ziconotide is still relatively limited, and the drug is currently used only in particularly severe cases of chronic pain, such as those that may occur in terminal cancer patients. However, intensive efforts are underway to develop synthetic forms of ziconotide that can independently reach the brain, in order to revolutionize the treatment of chronic pain.

A third major medical development from animal venoms is exenatide, trade name Byetta. This small peptide was isolated from the venom of the Gila monster (*Heloderma suspectum*). It shows great structural similarity to glucagon-like peptides that regulate insulin production in the human body. Exenatide and some of its derivatives have already been successfully used for some time to treat type II diabetes. Given the increasing life expectancy, a further increase in the use of this component can be expected in the future.

In addition to these three selected examples, there are other active substances derived from animal venoms, such as snake toxins for the treatment of blood clots or peripheral artery disease, leech venom for the treatment of coagulation disorders, or various chemically modified versions of captopril and exenatide. Even the raw venoms themselves are used. For example, the treatment of inflammation with bee venom (apitherapy) is indispensable in natural medicine and is also approved for this purpose. Similarly, leech therapy, or hirudotherapy, is frequently used in both human and veterinary medicine.

But animal venoms have not only been used in the treatment of diseases. They also play an important role in basic research and help to understand how molecules function in the body and how diseases develop. For example, they play a significant role in the characterization of receptors, particularly neurotransmitter receptors and ion channels. These proteins are responsible for transmitting signals between nerve cells and other cells in the body and are therefore crucial for numerous physiological processes. Animal venoms often contain a variety of bioactive molecules that specifically bind to these receptors and can modulate their function. By isolating and studying these toxins, researchers can gain important insights into the structure and function of these receptors. A commonly used method for characterizing receptors with animal venoms is the patch-clamp technique, which allows the measurement of the electrical activity of individual cells. Animal venoms serve as tools to manipulate the activity of specific receptors and to investigate their role in various physiological processes. For example, neurotoxins from snake venoms can be used to study the function of ion channels in the nervous system and to understand their involvement in diseases such as epilepsy or neuropathic pain. In addition, animal venoms can also serve as tools in structural biology to elucidate the three-dimensional structure of receptors. Some toxins bind highly specific to certain regions of receptors and can therefore be used as probes to identify their binding sites and to characterize structure-activity relationships. This enables researchers to develop detailed models of receptor structure and to identify potential target molecules for the development of new drugs. Overall, animal venoms play a crucial role in receptor research and make a significant contribution to our understanding of cellular signal transduction and the development of new pharmacological therapies.

New Perspectives with Animal Venoms

The study of animal toxins for medical applications has already provided modern medicine with several highly effective components that save lives every year. Despite the great potential hidden in animal venoms, it must be noted that only a few species have been thoroughly investigated.

In particular, the large venomous animals, especially reptiles, have been studied so far. However, these represent only a small fraction of the global diversity of venomous animals. The greatest species diversity is found among the smaller species, especially those belonging to the phylum Arthropoda, whose members and their venoms have so far received little attention. There is therefore still enormous potential for the discovery of new therapeutic agents among the many unexplored species of arthropods that are still waiting to be studied. In addition, the venoms of these small venomous animals often exhibit significantly higher chemical complexity. Spider venom can contain over 3,000 different toxins, whereas the toxin diversity of snake venoms is considerably lower by comparison. There are also about 20 times more venomous spiders than venomous snakes in nature. This extensive diversity of species and molecules means that the majority of components that can be isolated from animal venoms will be found in these neglected animal groups. Nevertheless, the venoms of precisely these species remain almost completely unexplored. This is particularly evident in spiders. It is estimated that around 10 million biomolecules may occur globally in spider venoms, which corresponds to about half of all toxins in the entire animal kingdom. However, of this enormous diversity, only a little more than 2,000 have been described, and well over 99% of all spider toxins remain unidentified. This imbalance is also reflected at the level of species diversity. Of the 52,000 spider species, only a few hundred have been studied with regard to their venoms. Similar ratios between unknown and known molecules, as well as between unstudied and studied species, exist for all smaller venomous animals, such as insects, pseudoscorpions, or venomous worms.

The widespread neglect of these smaller venomous animals is mainly due to the methodological approach of animal venom bioprospecting. Traditional methods of natural product research are primarily used here, in which the individual components are gradually isolated from crude venom by liquid chromatography and then examined individually for their function and structure. This approach is extremely time-consuming and costly, but above all, it is very sample-intensive. For a single chromatographic purification with subsequent functional tests of the components, several milligrams of crude venom may be required. Such quantities can be easily obtained from large venomous animals, but for smaller species, hundreds or even thousands of animals may need to be sampled. Often, venom sampling fails purely for physical reasons due to the small body size. For these reasons, small venomous animals have rarely been studied in the past, and their enormous potential for bioprospecting remains largely untapped. A remarkable achievement of recent years has been the development of new methods in chemical analytics, particularly mass spectrometry, which enables extremely precise detection even of the smallest sample quantities. With the increase in computing power and advances in bioinformatics, these highly precise methods can now be applied in high-throughput. These approaches from proteomics have revolutionized modern animal venom research and now make it possible to decipher the venoms of most species. Sequencing technologies have also become important pillars of animal venom research. In cases where venom extraction fails due to size, the venom glands of the animals can be surgically removed and the mRNA contained within sequenced. The resulting transcriptome (the entirety of all mRNA) then allows the venom composition of a species to be predicted by computer. Other important analytical innovations include techniques from metabolomics, which involve the analysis of secondary metabolites and small organic compounds, as well as the prediction of venom components from whole genomes. The use of mass spectrometric imaging technologies to determine the distribution of venom components within the venom glands also makes it possible to study the modulation and maturation of venoms in the respective species.

Although the analytical techniques mentioned above provide extremely effective solutions for deciphering venom compositions, they alone are not sufficient to generate new active substances, as they cannot provide bioactivity data. To achieve this goal, the individual pharmacologically interesting components must be produced in the laboratory. Since there is often not enough crude venom available, traditional isolation is not practical for these species. For this reason, biotechnology has recently become a key technology. Selected components previously identified using the methods mentioned above can now be specifically produced in bacterial cells on a large scale and subsequently tested. These new methods of animal venom research are collectively considered a new research field, internationally known as "Modern Venomics." Through Modern Venomics, theoretically all venom cocktails across the animal kingdom are now accessible for bioprospecting purposes. As a result, many previously unknown venoms have been identified in recent years, especially from small arthropods such as insects or arachnids, but also the venoms of sea anemones, fish, and worms have been explored using Modern Venomics. With the expansion of the range of species for drug discovery, some extremely promising drug candidates have already been identified.

One of the probably most promising components is Hi1a from the venom of the Australian funnel-web spider *Hadronyche infensa* (Fig. 3). Hi1a is a small peptide characterized by the presence of the ICK motif ("inhibitor cystine knot"), which consists of knot-like cross-linked disulfide bridges. This structure gives the component high stability against proteolytic degradation. Hi1a binds to "acid sensing ion channels" (ASIC), a class of receptors that play a significant role in cell death triggered by oxygen deprivation. In mouse models, it has been shown that Hi1a is capable of almost completely alleviating neuronal damage following ischemic strokes, even when the component is administered several hours after the stroke. Strokes represent a major medical challenge worldwide and are among the most common causes of disability and death. They occur when the blood supply to a part

Fig. 3 Funnel-web spider *Hadronyche infensa.* (© Bjoern Bartsch/Getty Images/iStock)

of the brain is interrupted, either by a blocked or ruptured blood vessel. This leads to a rapid loss of brain functions, including motor control, speech, and perception. Hi1a thus holds unexpected potential in combating one of humanity's most significant medical complications.

Interestingly, ASICs also play a crucial role in cell death in explanted organs. In animal models, Hi1a was also used to dramatically increase the viability and thus the transplantability of explanted hearts. It is therefore becoming apparent that Hi1a and its analogues could potentially be used both to combat neurological deficits after strokes and to extend the preservation time of rare hearts suitable for transplantation. These potential applications would represent a revolutionary advance in medicine.

In addition to the applications already discussed for stroke and heart transplantation, further highly promising biomedical applications are emerging, especially with spider venoms. In mouse models, a peptide from tarantula venom was successfully used to alleviate Dravet syndrome. Dravet syndrome is a rare and severe form of epilepsy that often begins in infancy. It is caused by mutations in the SCN1A gene, which encodes an important protein in the brain's nerve cells. People with Dravet syndrome typically suffer from seizures that are difficult to control and can take various forms, including generalized tonic-clonic seizures, and can even be fatal. In addition to seizures, developmental delays, behavioral problems, and motor impairments can also occur. Dravet syndrome is currently incurable. This underscores the potential of animal venoms in combating diseases that are difficult to treat. Further applications of animal venoms, particularly for the treatment of central nervous system disorders, are currently being intensively investigated.

Apart from cardiovascular and neurological diseases, applications in the field of anti-infectives (antimicrobial, antifungal, and antiviral components) are also conceivable. In particular, many such components are known from the venom of wolf spiders. Recently, our team was able to functionally investigate a series of wolf spider toxins that showed inhibition of the growth of pathogenic bacteria. Although no suitable components for human application have yet been identified, this work suggests that the molecular diversity in wolf spiders provides a promising basis for the search for new antimicrobial agents. In addition to wolf spiders, ant venoms have also been studied, with a small linear toxin discovered that showed promising activity against *Listeria monocytogenes*, a pathogen causing meningitis in newborns. Further research on this component is currently underway. However, a particularly exciting animal venom component in the context of the search for new anti-infectives is the checacins, a toxin family from the pseudoscorpion *Chelifer cancroides*. The toxin Checacin 1 shows highly potent activity against methicillin-resistant *Staphylococcus aureus*, a globally significant multi-resistant hospital pathogen, while its activity against human cells is low. There is an ongoing effort to further investigate the activities of Checacin 1 and, if appropriate, to develop this toxin further.

But animal venoms could also benefit humanity in the future beyond medicine. The use of spider venoms in agriculture, for example, represents a promising way to control pests effectively and in an environmentally friendly manner. Some spider venoms contain toxins that act specifically on insects and are harmless to other organisms, including cultivated plants and vertebrates. By targeted application of spider venoms, farmers could control pests such as aphids, caterpillars, and beetles without having to resort to harmful chemical pesticides. This alternative method could help reduce environmental pollution from pesticide use while improving yields and the quality of agricultural products. In addition, it can make agricultural work significantly safer. For although insecticides are indispensable for pest control in agriculture, they pose potential hazards to users. Many of these chemicals can be toxic and, if used improperly or through exposure, can present serious health risks to humans. Possible dangers include acute effects such as skin irritation, respiratory problems, or poisoning, as well as long-term effects such as chronic diseases or even cancer. In contrast, the vast majority of spider toxins are only minimally toxic to humans and thus represent a much safer alternative in crop protection.

A final application option for animal venoms can be found in industrial goods production. Here, enzymes have been used for decades to manufacture a variety of products. Animal venoms often also contain highly effective enzymes with interesting properties, and these could be used in the future to produce industrial goods in a sustainable way.

Conclusion

The investigation and study of animal venoms is undoubtedly an extremely fascinating and multifaceted field of research within applied zoology. This fascination arises from the remarkable duality of these venoms: they are not only the cause of suffering, but also a source of a multitude of beneficial molecules that could potentially have life-saving applications in biomedicine.

Despite the progress made so far and the extraction of some high-quality active substances from animal venoms in the past, it is important to note that a significant proportion of potentially interesting molecules remains undiscovered, as only a limited number of animal species have been thoroughly studied to date. However, thanks to recent advances in research methodology, especially in the field of modern venomics, entirely new opportunities are now opening up to tap into the entire animal kingdom for bioprospecting. Future work will focus intensively on deciphering these as yet undiscovered venoms and systematically harnessing the biomolecular treasures they contain. These promising developments could mean that the most potent weapons of the animal kingdom will ultimately make a decisive contribution not only to shaping medicine, but also to advancing agriculture and the production of goods in the future.

Conclusions

Communication between living organisms by means of chemical signals began long before humans appeared on Earth. Chemical communication has accompanied the millions of years of evolution of life on our planet and forms the basis for the enduring ecological connections between organisms. The highly sensitive and delicate channel of chemical communication helps maintain balance and harmony in nature. With the emergence of humans and their establishment as the dominant biological species on the planet, this balance began to change. With the onset of the scientific and technological revolution, humans began to intervene massively in the life of the biosphere. In the last 100–200 years, we have witnessed changes on the Earth's surface that are equivalent to the greatest upheavals over millions of years. Unfortunately, humanity's sense of gratitude, respect, and responsibility toward nature has developed much more slowly than technological progress. For many years, humans viewed nature as an adversary to be fought bitterly and relentlessly. For a long time, we saw the achievements of technological progress as new means to fight and subjugate nature. It took a long time for us to recognize our belonging to nature and the fact that any activity directed against nature is also an activity against humanity itself. As a result of irresponsible human actions, numerous plant and animal species have disappeared from the face of the earth forever. But technological progress is a fact, and it is the foundation for the rapid social development of humanity. Like every other human achievement, this progress has its good and bad sides. We know that a stone picked up from the road and tied to a stick can be both a tool for work and a weapon of destruction. In the same way, the

I. G. Ivanov, *The Invisible Language of Nature*,
https://doi.org/10.1007/978-3-662-73302-8_44

achievements of technological progress are constructive in one respect, but destructive in another.

The rapid development of ecological science and its popularization in recent decades have brought important insights into the complex relationships between living beings on Earth on the one hand, and between them and humans on the other. Nature, too, has been granted a right to protection. However, the law can only protect what is objectively proven to be in need of protection. Despite the successes of ecological science, it is still not able to foresee all the long-term consequences of one or another human activity on nature and its balance. Here is a simple example. What would happen if a fire broke out somewhere in nature? Everyone would answer: "That depends on the size and location of the fire." The quantitative damage of a fire is measured in hectares of destroyed forest, tons of hay, straw, oil, etc. However, no one talks about the animals destroyed that live in the affected areas, as they are not recorded in the news reports. One might assume that their number is not very large, since many of them will escape as the fire spreads. But exactly the opposite was observed in a major fire in California. It broke out in an oil refinery and destroyed 120,000 t of oil in a short time. Completely unexpectedly, countless swarms of jewel beetles (Buprestidae) of the species *Melanophila consputa* and *Melanophila atropurpurea* made their way into the fire and burned. These beetles live in juniper stands. It turned out that some of the components of the smoke (it is not exactly known which) are related to their aggregation pheromones. Interestingly, in this case, the nearest juniper forest from which they could have originated was 80 km from the site of the fire.

Today, the harmfulness of industrial waste released into the environment is mainly assessed based on its toxicity. However, chemical ecology, which is briefly introduced in this book, teaches us that toxicity is by no means a prerequisite for a substance to have a harmful biological effect. Chemical compounds of the pheromone type, which exert their effect via the olfactory system, are not toxic, but can nevertheless lead to significant changes in the life cycle of the organisms exposed to them. We have seen that the sexual attractant for the New Zealand scarab beetle species *Costelytra zealandica* is phenol. That seems strange to us today. Phenol can be found under, in, and on the street. Phenol solutions are used to disinfect public vehicles and sanitary facilities in public buildings. Phenol is an important raw material for the production of plastics, paints, varnishes, pharmaceuticals, explosives, etc. Before the era of large-scale chemistry, however, phenol was as unique in nature as the atomic bomb is today. So we have no reason to blame the New Zealand beetle for choosing phenol as its sex pheromone. It simply did not know that the time would come when humans would begin to release tons of phenol into the atmosphere, thereby preventing male beetles from finding a mate. Unfortunately, phenol is not the only pheromone that the chemical industry produces as a competing substance. Closely related to pheromones are the numerous substances released into the air and water by the oil, forestry, perfume, and other industries. According to their product properties, these wastes are non-toxic and are therefore considered harmless. But does anyone consider their effects on the reproductive capacity of animals? The suppression of the reproductive process due to difficulties in finding a mate is no less harmful to a species than its physical extermination. Numerous detergents are considered non-toxic. However, when they are discharged into rivers, they destroy the cilia of the olfactory epithelium of fish, which is equivalent to blindness for them.

Advances in the study of chemical relationships between animals are making it increasingly clear that the term "harmful substance" must not be equated with "toxic substance." A substance may be completely harmless to the individual, but dangerous to the species. This is what chemical ecology teaches us, which in recent years has provided us with valuable insights into the complex interrelationships between living beings. These insights are essential for the wise management of technological progress and for an appropriate resolution of the human-nature conflict of our time.

Glossary

Allomone Signaling substances that transmit information between individuals of different species. The information is exclusively advantageous for the sender.

Angiosperms Flowering plants, angiosperms, the largest class of seed plants; their ovules are enclosed and protected by a carpel or ovary ("covered").

Anosmia complete loss of the sense of smell

Antheridia male gametangium (sexual organ) in, for example, mosses, ferns, club-mosses, certain algae, and fungi

Aphrodisiacs Active substances that stimulate or increase libido

Arthropods Arthropods; invertebrate animals such as spiders, crustaceans, millipedes, and insects

Autotrophic Ability of organisms to build their structural components exclusively from inorganic substances

Bioprospecting Assessment of the commercial potential of biological resources, especially for medical purposes

Chemoreception Physiological process in which chemical signals from the environment are converted into an action potential via corresponding receptors and become assignable for the CNS, such as smell or taste

Chemotaxis Influence of the direction of movement of organisms or cells by a concentration gradient of an active substance

Chitin Horn-like main component of the exoskeleton of crustaceans, insects, spiders, and millipedes

cis-trans isomerism Specification of whether molecules differ in that two substituents are on the same side of a reference plane or not

Chromophores Dye in which excitable electrons are present

I. Ivanov, *The Invisible Language of Nature*, https://doi.org/10.1007/978-3-662-73302-8

Dalton, kiloDalton (kDa) Name for the atomic mass unit, exactly equal to 1/12 the mass of the carbon isotope ^{12}C and corresponds approximately to the mass of a hydrogen atom

Denaturation Structural change of, for example, proteins with a loss of biological function, while the primary structure remains unchanged

DNA Deoxyribonucleic acid ("deoxyribonucleic acid"), carries genetic information in all living organisms

Dystrophic malnourished

Eccrine secreting outward

Ethology Science of animal and human behavior

Eukaryotes Organisms whose cells possess a nucleus

Exocrine see also **eccrine;** secretion onto external (e.g., sweat via the skin) or internal surfaces (e.g., saliva into the oral cavity, digestive enzymes into the intestine)

extracellular matrix Part of the tissue that lies between the cells and surrounds them in a mesh-like manner

Gametes collective term for sperm (pollen) and egg cells

Gamones Pheromones (see there) of gametes

Gustatory relating to the sense of taste, tasting

Gymnosperms Seed plants whose seeds are not surrounded by fruit flesh

Hemolymph Body fluid of invertebrate animals without a closed circulatory system

Heterotrophic Organism that cannot produce its own food; obtains nutrients from other sources of organic carbon, plant or animal

Homo sapiens Representative of the genus Homo (humans)

Hormone Hormones are chemical messengers produced in the body to regulate important bodily functions.

Hydrolysis Splitting of chemical compounds by a reaction with water

Hydrophobic Substances that are not or only poorly soluble in water

in vitro scientific experiment conducted in a test tube

ionophoric Molecule capable of binding charged ions (cations or anions)

Kairomones Messenger substance for information transfer between different species; benefits only the receiving organism, the recipient

kDa see Dalton

convergent approaching each other, coinciding

Lipophilicity Substances with a tendency to associate with the molecules of fats or oils ("fat-loving")

Metabolome encompasses all characteristic metabolic properties of a cell, tissue, or organism

Microbiome Totality of all microorganisms (e.g., bacteria or viruses) that colonize a living organism

mcg/ml Micrograms per milliliter

Musk Musk, also called civet, is a strongly scented secretion of the male musk deer

mRNA ("messenger ribonucleic acid"), messenger RNA; transmits genetic information for the synthesis of a specific protein in a cell

Mycelium filamentous cells (hyphae) of fungi, which spread as a branching fungal network

Nematode Roundworms, a species-rich phylum of the animal kingdom, with more than 20,000 known species

Neostigmine reversible cholinesterase inhibitor, inhibits the breakdown of acetylcholine

Odorology Science of odor, used as a branch of criminology and forensics for the identification of individuals by their unique scent

Ecology Science of the interactions between living organisms and their environment

Olfactory relating to the sense of smell

Ontogenetic development of an individual or a single organism

Osmophores characteristic functional groups found in odorants, often pleasantly scented

Pheromones secreted scent substance that influences the metabolism and behavior of other individuals of the same species

Phylogenetic evolutionary development of all living organisms and their related groups

Phytoalexin antimicrobial substances produced by plants to defend against pathogenic germs

Phytoncides substances produced by a plant that inhibit the growth of or are lethal to pathogens

Primates Order that includes prosimians, monkeys, apes, and thus also humans

Olfactory bulb Bulbus olfactorius, a swelling at the anterior base of the brain

RNA single strand built from nucleotides in every cell of a living organism, important carrier of information and function in a cell

Saprophytic heterotrophic mode of nutrition in which dead organic material serves as substrate

Semiochemicals Messenger substances that serve chemical communication between individuals of a species or between different species.

Speciation the formation of several species from one species through reproductive isolation of subpopulations with different gene pools

Spores serve asexual reproduction, are very resistant, and can completely shut down their metabolism

Symbiosis Association of individuals of two different species that is advantageous for both partners

Synomones Signaling substances that transmit information between individuals of different species

Taxonomy Systematics for classifying living organisms into categories

Toxin Poison synthesized by a living organism

Transcriptome Sum of all genes transcribed in a cell at a given time, i.e., copied from DNA into RNA (see there)

Trophallaxis zoological term for the transfer of liquid food from the mouth or anus of one animal to another

Van der Waals forces Attractive forces between two molecules that spontaneously develop dipoles. Due to the unequal distribution of charge, the oppositely charged regions of the molecules attract each other.

Further Reading

Abd El-Ghany NM (2019) Semiochemicals for controlling insect pests (Review). J Plant Prot Res 59:1–11. https://doi.org/10.24425/jppr.2019.126036

Achyuthan VS, Sunagar SS (2018) Animal venoms: origin, diversity and evolution. Wiley Online Library. https://doi.org/10.1002/9780470015902.a0000939.pub2

Ada FB, Sunday KI, Ugbong EA (2021) Animal venoms (Book). GSC Biol Pharm Sci 14:047–054. https://doi.org/10.30574/gscbps.2021.14.1.0371

Aldich JR (1995) Chemical communication in the true bugs and parasitoid exploitation. In: Carde RT, Bell WJ (eds) Chemical ecology of insects. Chapman & Hall, New York, pp 318–363

Aldrich JR, Oliver JE, Taghizadeh T, Ferreira JTB, Liewehr D (1999) Pheromones and colonization: reassessment of the milkweed bug migration model (Heteroptera: Lygaeidae; Lygaeinae). Chemoecology 9:63–71

Aldrich JR, Khrimian A, Zhang A, Shearer PW (2006) Bug pheromones (Hemiptera, Heteroptera) and tachinid fly host- finding. Denisia 19, zugleich Kataloge der OÖ. Landesmuseen Neue Serie 50:1015–1031

Alidrich JR, Rosi MC, Bin F (1995) Behavioral correlates for minor volatile compounds from stink bugs (Heteroptera: Pentatomidae). J Chem Ecol 21:1907–1920

Arnold G, Le Conte Y, Trouiller J, Hervet H, Chappe B, Masson C (1994) Inhibition of worker honeybee ovaries development by a mixture of fatty acid esters from larvae. C R Acad Sci, Ser 3 Sci vie 317: 511–515

Avitabile AR, Morse RE, Boch R (1975) Swarming honeybees guided by pheromones. Ann Entomol Soc Am 68:1079–1082

Barbier J, Lederer E (1960) Structure chimique de la substance royale de la reine d'abeille (Apis mellifera L.). C R Acad Sci, Ser 3 Sci vie 251:1131–1135

Benelli G, Lucchi D (2021) From insect pheromones to mating disruption: theory and practice. Insects 12:698

I. Ivanov, *The Invisible Language of Nature*, https://doi.org/10.1007/978-3-662-73302-8

Blum MS, Brand JA (1972) Social insect pheromones: their chemistry and function. Am Zool 12:553–576

Breed MD, Stiller TM, Blum MS, Page RE (1992) Honeybee nestmate recognition – effects of queen fecal pheromones. J Chem Ecol 18:1633–1640

Breed MD, Garry MF, Pearce AN, Hibbard BE, Bjostad LB, Page RE (1995) The role of wax comb in honey bee nestmate recognition. Anim Behav 50:489–496

Breed MD, Leger EA, Pearce AN, Wang YJ (1998) Comb wax effects on the ontogeny of honey bee nestmate recognition. Anim Behav 55:13–20

Breed MD, Guzman-Novoa E, Hunt GJ (2004) Defensive behavior of honey bees: organization, genetics, and comparisons with other bees. Annu Rev Entomol 49:271–298

Brillet C, Robinson GE, Bues R, Le Conte Y (2002) Racial differences in division of labor in colonies of the honey bee (Apis mellifera). Ethology 108:115–126

Butler CG, Fairey EM (1963) The role of the queen in preventing oogenesis in worker honey bees. J Apic Res 21:14–18

Butler CG, Callow RK, Johnston NC (1961) The isolation and synthesis of queen substance, 9-oxodec-trans-2-enoic acid, a honeybee pheromone. Proc R Soc Lond B Biol Sci 155:417–432

Carde RT, Backer TC (1984) Sexual communications with pheromones, Chapter 13. In: Bell WJ, Carde RT (eds) Chemical ecology of insects. Springer, S 355–383, ISBN 978- 0-412-23262-2

Chang CC, Lee CY (1963) Isolation of neurotoxins from the venom of bungarus multicinctus and their modes of neuromuscular blocking action. Arch Int Pharmacodyn Ther 144:241–257

Chen X, Gottelieb L, Millar JG (2000) Highly stereoselective syntheses of the sex pheromone components of the southern green stink bug Nezara viridula L. and the green stink bug Acrosternum hilare (SAY). Synthesis 2:269–272

Conte Le Y, Arnold G, Trouiller J, Masson C, Chappe B, Ourisson G (1989) Attraction of the parasitic mite Varroa to the drone larvae of honey bees by simple aliphatic esters. Science 245:638–639

Conte Le Y, Arnold G, Trouiller J, Masson C, Chappe B (1990) Identification of a brood pheromone in honeybees. Naturwissenschaften 77:334–336

Conte Le Y, Sreng L, Trouiller J (1994) The recognition of larvae by worker honeybees. Naturwissenschaften 81:462–465

Conte Le Y, Sreng L, Poitout SH (1995) Brood pheromone can modulate the feeding behavior of Apis mellifera workers (Hymenoptera: Apidae). J Econ Entomol 88:798–804

Conte Le Y, Mohammedi A, Robinson GE (2001) Primer effects of a brood pheromone on honeybee behavioural development. Proc R Soc Lond B Biol Sci 268:163–168

Deisig N, Dupuy F, Anton S, Renou M (2014) Responses to pheromones in a complex odor world: sensory processing and behavior (Review). Insects 5:399–422

Fitzgerald TD, St. Clair AD, Daterman GE, Smith RG (1973) Slow release plastic formulation of the cabbage looper pheromone cis-7-dodecenyl acetate: release rate and biological activity. Environ Entomol 2:607–610

Frisch von K (1967) The dance language and orientation of bees. Harvard University Press, Cambridge, MA

Gary NE (1962) Chemical mating attractants in the queen honey bee. Science 136:773–774

Groot AP, Voogd A (1954) On the ovary development in queenless worker bees (Apis mellifera L.). Experientia 10:384–385

Grozinger CM, Sharabash NM, Whitfield CW, Robinson GE (2003) Pheromone-mediated gene expression in the honey bee brain. Proc Natl Acad Sci USA 100:14519–14525

Haddad V Jr, Amorim PCH, Haddad WT Jr, Cardoso JLC (2015) Venomous and poisonous arthropods: identification, clinical manifestations of envenomation, and treatments used in human injuries. Rev Soc Bras Med Trop 48:650–657

Higo HA, Winston ML, Slessor KN (1992) Mechanisms by which honey bee (Hymenoptera: Apidae) queen pheromone sprays enhance pollination. Ann Entomol Soc Am 88:366–373

Hoover SER, Keeling CI, Winston ML, Slessor KN (2003) The effect of queen pheromones on worker honey bee ovary development. Naturwissenschaften 90:477–480

Huang ZY, Robinson GE (1996) Regulation of honey bee division of labor by colony age demography. Behav Ecol Sociobiol 39:147–158

Hunt GJ, Wood KV, Guzman-Novoa E, Lee HD, Rothwell AP, Bonham CC (2003) Discovery of 3-methyl-2-buten-yl acetate, a new alarm component in the sting apparatus of Africanized honeybees. J Chem Ecol 29:453–463

Jay SC (1968) Factors influencing ovary development of worker honeybees under natural conditions. Can J Zool 46:345–347

Jay SC (1972) Ovarian development of worker honeybees when separated from worker brood by various methods. Can J Zool 50:661–664

John B, Watkins III (2020) Toxic effects of terrestrial animal venoms and poisons. Chapter 26. In: Home books Casarett & Doull's essentials of toxicology (Book), 2nd edn. McGraw-Hill,

Junghanss T, Bodio M. (2006) Medically important venomous animals: biology, prevention, first aid, and clinical management. Clin Infect Dis 43(10):1309–1317

Junior VH, Neto JBP, Cobo VJ (2006) Venomous mollusks: the risks of human accidents by Conus snails (Gastropoda: Conidae) in Brazil. Rev Soc Bras Med Trop 39:498–500

Kaatz HH, Hildebrandt H, Engels W (1992) Primer effect of queen pheromone on juvenile hormone biosynthesis in adult worker honey bees. J Comp Physiol B 162:588–592

Kainoh Y, Tanaka C, Nakamura S (1999) Odor from herbivoredamaged plant attracts the parasitoid fly Exorista japonica (Diptera: Tachinidae). Appl Entomol Zool 34:463–467

Kalia J, Milescu M, Salvatierra J, Wagner J, Klint JK, King GF, Olivera M, Bosmans F (2015) From foe to friend: using animal toxins to investigate ion channel function. J Mol Biol 427:158–175

Katzav-Gozansky T, Soroker V, Hefetz A, Cojocaru M, Erdmann DH, Francke W (1997) Plasticity of caste-specific Dufour's gland secretion in the honey bee (Apis mellifera L.). Naturwissenschaften 84:238–241

Katzav-Gozansky T, Soroker V, Ionescu A, Robinson GE, Hefetz A (2001a) Task-related chemical analysis of labial gland volatile secretion in worker honeybees (Apis mellifera). J Chem Ecol 27:919–926

Katzav-Gozansky T, Soroker V, Ibarra F, Francke W, Hefetz A (2001b) Dufour's gland secretion of the queen honeybee (Apis mellifera): an egg discriminator pheromone or a queen signal? Behav Ecol Sociobiol 51:76–86

Keeling CI (2001) Isolation and identification of new components of the honey bee (Apis mellifera) queen retinue pheromone. PhD dissertation, Simon Fraser University, Vancouver

Keeling CI, Slessor KN, Higo HA, Winston ML (2003) New components of the honey bee (Apis mellifera L.) queen retinue pheromone. Proc Natl Acad Sci USA 100:4486–4491

Khrimian A (2005) The geometric isomers of methyl-2,4,6- decatrienoate, including pheromones of at least two species of stink bugs. Tetrahedron 61:3651–3657

Kochansky J, Aldrich JR, Lusby WR (1989) Synthesis and pheromonal activity of 6,10,13-trimethyl-1-tetradecanol for the predatory stink bug Stiretrus anchorago (Heteroptera: Pentatomidae). J Chem Ecol 15:1717–1728

Laminna C (1959) The most poisonous poison. Science 130:763–772

Ledoux MN, Winston ML, Higo H, Keeling CI, Slessor KN, Conte LY (2001) Queen pheromonal factors influencing comb construction by simulated honey bee (Apis mellifera L.) swarms. Insect Soc 48:14–20

Leoncini I (2002) Phéromones et régulation sociale chez l'abeille Apis mellifera L.: Identification d'un inhibiteur du développement comportemental des ouvrières. PhD. dissertation, Institut National Agronomique Paris-Grignon, Paris

Leoncini I, Conte Le Y, Costagliola G, Plettner E, Toth AL, Wang M, Huang Z, Bécard J, Crauser D, Slessor KN, Robinson GE (2004) Regulation of behavioral maturation by a primer pheromone produced by adult worker honey bees. Proc Natl Acad Sci USA 101:17559–17564

Martin SJ, Jones GR (2004) Conservation of biosynthetic pheromone pathways in honeybees. Apis Naturwissenschaften 91:232–236

Mata da ÉCG, Mourão CBF, Rangel M, Schwartz EF (2017) Antiviral activity of animal venom peptides and related compounds. J Venom Anim Toxins Incl Tropl Dis 23:3–15

McBrien HL, Millar JG (1999) Phytophagous bugs. In: Hardie J, Minks AK (eds) Pheromones of non-lepidopteran insects associated with agricultural plants. CAB International Publishing, Wallingford, pp 277–304

McBrien HL, Millar JG, Gottlieb L, Chen X, Rice RE (2001) Male-produced sex attractant pheromone of the green stink bug Acrosternum hilare (SAY). J Chem Ecol 27:1821–1839

McBrien HL, Millar JG, Rice RE, Mccelfresh JS, Cullen E, Zalom FG (2002) Sex attractant pheromone of the redshouldered stink bug Thyanta pallidovirens: a pheromone blend with multiple redundant components. J Chem Ecol 28:1797–1818

Meier J (2008) Handbook of: clinical toxicology of animal venoms and poisons, 1st edn. CRC Press. https://doi.org/10.1201/9780203719442

Michener CD (1969) Comparative social behavior of bees. Annu Rev Entomol 14:299–342

Mohammedi A, Crauser D, Paris A, Conte LY (1996) Effect of a brood pheromone on honeybee hypopharyngeal glands. C R Acad Sci Paris, Sci Vie/Life Sci 319:769–772

Mohammedi A, Paris A, Crauser D, Conte LY (1998) Effect of aliphatic esters on ovary development of queenless bees (Apis mellifera L.). Naturwissenschaften 85:455–458

Naumann K, Winston ML, Slessor KN, Prestwich GD, Webster FX (1991) Production and transmission of honey bee queen (Apis mellifera L.) mandibular gland pheromone. Behav Ecol Sociobiol 29:321–332

Naumann K, Winston ML, Slessor KN (1993) Movement of honey bee (Apis mellifera L.) queen mandibular gland pheromone in populous and unpopulous colonies. J Insect Behav 6:211–223

Pain J (1961) Sur la phéromone des reines d'abeilles et ses effets physiologiques. Ann Abeille 4:73–152

Pankiw T (2004) Cued in: honey bee pheromones as information flow and collective decision-making. Apidologie 35:217–226

Pankiw T, Page RE (2001) Brood pheromone modulates honeybee (Apis mellifera L.) sucrose response thresholds. Behav Ecol Sociobiol 49:206–213

Pankiw T, Huang ZY, Winston ML, Robinson GE (1998) Queen mandibular gland pheromone influences worker honey bee (Apis mellifera L.) foraging ontogeny and juvenile hormone titers. J Insect Physiol 44:685–692

Pankiw T, Winston ML, Fondrk MK, Slessor KN (2000) Selection on worker honeybee responses to queen pheromone (Apis mellifera L.). Naturwissenschaften 87:487–490

Pankiw T, Roman R, Sagili R, Zhu-Salzman K (2004) Pheromone-modulated behavioral suites influence colony growth in the honey bee (Apis mellifera). Naturwissenschaften 91:575–578

Pickett JA, Williams IH, Smith MC, Martin AP (1980) Part I. Chemical characterization. J Chem Ecol 6:425–434

Plettner E, Slessor KN, Winston ML, Oliver JE (1996) Caste-selective pheromone biosynthesis in honeybees. Science 271:1851–1853

Pozio E (1988) Venomous snake bite in Italy: epidermalogy and clinical aspects. Trop Med Parasitol 39:62–66

Rajchard J (2005) Sex pheromones in amphibians: a review. Vet Med Czech 50:385–389

Ratnieks FLW (1995) Evidence for a queen-produced egg-marking pheromone and its use in worker policing in the honey bee. J Apic Res 34:31–37

Ratnieks FLW, Visscher PK (1989) Workers policing in the honeybee. Nature 342:796–797

Regnier FE, Law JH (1968) Insect pheromones (Review). J Lipid Res 9:542–550

Robinson GE (1992) Regulation of division of labor in insect societies. Annu Rev Entomol 37:637–665

Robinson GE, Page RE Jr, Strambi C, Strambi A (1989) Hormonal and genetic control of behavioral integration in honey bee colonies. Science 246:109–112

Robinson GE, Fernald RD, Clayton D (2008) Genes and social behavior. Science 322:896–900

Rong MA, Rangel J, Grozinger CM (2019) Honey bee (Apis mellifera) larval pheromones may regulate gene expression related to foraging task specialization. BMC Genomics 20:592–607

Sabatier JM, Cao Z, Wang JL, McNutt PM, Utkin YN, Kovacic H, Shahbazzadeh D, Wulff H (eds) (2021) Venom, animal and microbial toxins. Frontiers Media SA, Lausanne. https://doi.org/10.3389/978-2-88966-991-2

Seeley TD (1979) Queen substance dispersal by messenger workers in honeybee colonies. Behav Ecol Sociobiol 5:391–415

Slessor KN, Higashi S (eds) (2003) Genes, behaviors and evolution of social insects. Hokkaido University Press, Sapporo, pp 55–77

Slessor KN, Kaminski LA, King GGS, Borden JH, Winston ML (1988) Semiochemical basis of the retinue response to queen honey bees. Nature 332:354–356

Slessor KN, Winston ML, Le Conte Y (2005) Pheromone communication in the honeybee (Apis mellifera L.). J Chem Ecol 31:2731–2745

Suranse V, Srikanthan A, Sunagar K (2018) Animal venoms: origin, diversity and evolution (Book). Wiley on line library: https://doi.org/10.1002/9780470015902.a0000939.pub2

Sutherland SK, Tibballs J (2001) Australian animal toxins: the creatures, their toxins and care of the poisoned patient, 2nd edn. Oxford University Press, Melbourne

Takács Z, Nathan S (2014) Animal venoms in medicine (Book). Semantic Scholars Publication. https://doi.org/10.1016/B978-0-12-386454-3.01241-0. Corpus ID: 81683809

Tomida I (1968) Pheromone and its application to agricultural chemicals. Japan Agricultural Research Quarterly, ISSN 00213551

Utkin YN (2015) Animal venom studies: current benefits and future developments. J Biol Chem 6:28–33

Whitfield CW, Band MR, Bonaldo MF, Kumar CG, Liu L, Pardinas JR, Robertson HM, Soares MB, Robinson GE (2002) Annotated expressed sequence tags and cDNA microarrays for studies of brain and behavior in the honey bee. Genome Res 12:555–566

Wilson EO, Bossert WH (1963) Chemical communication among animals. Recent Prog Horm Res 19:673–716

Winston ML (1987) The biology of the honey bee. Harvard University Press, Cambridge, MA

Winston ML, Higo HA, Colley SJ, Pankiw T, Slessor KN (1991) The role of queen mandibular pheromone and colony congestion in honey bee (Apis mellifera L.) reproductive swarming. J Insect Behav 4:649–660

Wossler TC, Crewe RM (1999a) Honey bee races (Apis mellifera). J Apic Res 38:137–148

Wossler TC, Crewe RM (1999b) The releaser effects of the tergal gland secretion of queen honeybees (Apis mellifera). J Insect Behav 12:343–351

Wyatt TD (2003) Pheromones and animal behavior (Book). Cambridge University Press